The Prentice Hall Guide to Research Writing

SECOND EDITION

The Prentice Hall Guide to Research Writing

Dean Memering

Central Michigan University

PRENTICE HALL, Upper Saddle River, NJ 07458

Library of Congress Cataloging-in-Publication Data

Memering, Dean, (date)
The Prentice Hall guide to research writing / Dean Memering.—
2nd ed.
p. cm.
First ed. published as: Research writing. c1983.
Includes index.
ISBN 0-13-774480-3
1. Report writing. 2. Reseach. I. Memering, Dean, (date)
Research writing. II. Title. III. Title: Guide to research
writing.
LB2369.M39 1989
808'.02—dc19 88-31603
CIP

Editorial/production supervision: Joseph O'Donnell
Interior and cover design: Meryl Poweski
Manufacturing buyer: Laura Crossland

A Pearson Education Company
Upper Saddle River, NJ 07458

Printed in the United States of America
10 9 8 7 6

ISBN 0-13-774480-3 01

Prentice-Hall International (UK) Limited,London
Prentice-Hall of Australia Pty. Limited, Sydney
Prentice-Hall Canada Inc., Toronto
Prentice-Hall Hispanoamericana, S.A., Mexico
Prentice-Hall of India Private Limited, New Delhi
Prentice-Hall of Japan, Inc., Tokyo
Pearson Education Asia Pte. Ltd., Singapore
Editora Prentice-Hall do Brasil, Ltda., Rio de Janeiro

Contents

CHAPTER TWO
USING THE LIBRARY 25

CHAPTER THREE
READING RESEARCH 61

CHAPTER FOUR
EVALUATING EVIDENCE 111

CHAPTER SIX
ORGANIZING RESEARCH WRITING *205*

CHAPTER SEVEN
STYLE IN RESEARCH WRITING *245*

APPENDIX
WRITING SITUATIONS *263*

Preface

We have come a long way since the days when teachers and textbooks asked students to write "term papers." The old tradition treated such papers as a mere matter of form and usually assumed that doing them was so simple that students needed very little instruction or help. Often the "term paper"—usually a long and boring collection of cribbed material—was simply tacked onto a course having little to do with term papers (i.e., any course might require such a paper: history, literature, geography, etc.). It was seldom thought worth spending much time on, so publishers put out little term-paper books and booklets, and the whole operation was treated fairly mechanically. Since it all seemed pointless and quite difficult to students, their papers were often incredibly boring and unimaginative, often plagiarized. Term-paper companies sprang up to take advantage of the fact that students needed these papers but really were not equipped to write them.

Today many teachers are dealing more realistically with this kind of writing, exploring its complexities and varieties. From being the simple-minded, pointless orphan of the curriculum, research writing is beginning to gain a much more important position. It is the kind of writing on which education is based, the kind of writing teachers and professors write and read. Indeed, it is the kind of writing upon which much of modern knowledge is based: the natural sciences, social sciences, industry, medicine, government—it is the heart of modern knowledge. Today, when we say we "know" something, we often mean we believe something we have read or heard. One of the achievements of the modern world has been to give the common man access to knowledge. Through the techniques of research we can each investigate and find out for ourselves what we need to know. Through the techniques of research writing, we can each compile research, put forth information, and express conclusions that others will believe.

The second edition of *Research Writing* takes a large step forward from the first edition: the book has further evolved away from the standard term-paper guides profusely available in college bookstores, toward a genuine textbook that teaches students how to do research. The notion that term papers or library papers merely add documentation to essay writing is an unrealistic approach to this specialized and difficult kind of writing. As increasing numbers of schools add a course in research writing, we must all adopt a more realistic appreciation for the challenge of writing with sources. Patching together a string of quotes cribbed from dubious sources is not research writing. Students must develop the mind set and investigatory habits of the scholar-researcher.

Writing with source material, using documentation, is an important part of the academic and working worlds; the modern world is more indebted to documented writing than most people know. The miracle of modern knowledge is our ability to pass on *verifiable* information. It is the *research* in research writing that provides the intellectual excitement and integrity of learning. When students learn the techniques of learning, the methods of researching, they will become independent learners: students can become experts; they can discover new information, develop new ideas, add to the world's store of knowledge. And it is the *writing* in research writing that gives form and structure to research. Our ability to write absolutely influences what we are able to write: how we write determines what we write. The written word makes knowledge visible and transmissible. Because so much depends on the accuracy of research writing, we are all necessarily vitally interested in the skills and strategies of writing and of the interaction between writing and researching. The two are inseparable.

Students must become at ease with the concept of substantiation. They must develop the intellectual curiosity that will lead them to pursue ideas, investigate side issues, account for contradictions, track down obscure references. It is not the one-shot ordeal of the long paper at semester's end that is important; it is the habit of working with sources, interpreting evidence, evaluating data, collecting and synthesizing information—in short, it is becoming a researcher that is important. And surely this is as difficult for students as becoming a poet or a composer or an accountant.

Students need practice, and guidance, and feedback. The dozens of term-paper books already available try to outdo each other in producing the longest list of what's in the library, in producing the most comprehensive set of obscure reference models, in producing the most precise and trivial typing guidelines. The term-paper books all have in common the guiding principle that the appearance of the paper is the only important consideration. Such books have nothing to say about the substance of research. By contrast, the second edition of *Research Writing* goes a long way toward establishing what it means to be a researcher today.

The book has two major divisions. The first part contains the first seven chapters, illustrating principles of research writing. The Appendix presents model papers with guidelines for writing for different purposes and audiences.

Chapter 1, The Research Process, introduces the basic concepts and attitudes of modern research writing. Research is presented as a series of cycles, rather than steps, through which the research moves: analyzing the research situation, beginning the preliminary reading, framing the research question, starting data collection, evaluating evidence, writing the paper. Chapter 1 devotes special attention to the key task of formulating the research question.

Chapter 2, Using the Library, explains the research features of modern college libraries. Important keys to the library, such as the card catalog, the indexes, and the general reference works, are illustrated in detail. New library machines are explained: computer terminals, ERIC, the microfilm indexes and microform readers, and new on-line databases like InfoTrac. In this second edition, Chapter Two has been organized around the library research strategy.

Chapter 3, Reading Research, is a new chapter. It presents techniques of reading for different purposes: reading for facts, reading for analysis, reading for interpretation, and reading for evaluation. The chapter explains the difference between active and passive reading and gives students techniques for interacting with reading material. Chapter 3 also explains how to take research notes.

Chapter 4, Evaluating Evidence, is a key chapter (along with Chapter 3). Students are shown how to think about data, how to analyze, interpret, and evaluate information. Rules of evidence and how to build a case with evidence make up the heart of the chapter: students learn the difference between primary and secondary evidence. The chapter also includes a complete presentation of the inductive fallacies—problems to avoid in research.

Chapter 5, Documentation, examines the problems of documentation from the student's point of view: how much documentation is required, what to document, and what not to document. The chapter gives detailed illustrations of the new MLA and APA note forms, as well as the traditional footnote/endnote style.

Chapter 6, Organizing Research Writing, presents a step-by-step development of the research paper, from the formal outline to the bibliography. Every section of the research paper has its own explanation and illustration.

Chapter 7, Style in Research Writing, illustrates the stylistic virtues of clarity, accuracy, conciseness. The chapter teaches students how to deal with problems such as passive mode, abstract language, vague language, sexist language, and inappropriate verb tenses in research writing. This chapter contains a brief guide to formal punctuation and capitalization.

The Appendix presents models and guidelines for various sorts of writing assignments students may face in a research writing class. The models include summaries, abstracts, analytical writing, writing a critique, writing a comparison, writing a review, writing a report, writing an argument, and writing a research paper.

ACKNOWLEDGMENTS

I am pleased to acknowledge the following colleagues who offered helpful advice during the preparation of the second edition: Donna Kay Dell, Devry Institute of Technology, and Peter Burton Ross, The University of the District of Columbia.

Special thanks are due to Beth McLeod, research librarian at Park Library, Central Michigan University, for her generous assistance with the new material in Chapter 2. And I am truly grateful for her help in collecting the sources on apes and language.

I owe a special debt to my wife, Joan, not only because she patiently endures the many hours I spend working on manuscript, but also for the insightful feedback she gives me on everything I write. A talented teacher in her own right, she is also my most helpful reader.

The Prentice Hall Guide to Research Writing

CHAPTER ONE

The Research Process

BECOMING A RESEARCH WRITER

Recently a friend came to me with a problem. He was preparing to apply for a red belt in karate, and as part of his application he was required to write a long paper about the sport of karate, the human body, and his values and goals in karate. Organizing the information and writing the report, he said, was becoming a problem, and would I be willing to review his paper? I know next to nothing about karate, but I agreed to help him. Much of the information for the paper he had collected firsthand from his karate lessons, some of it came from books on karate, and some came from doctors and other experts my friend had talked to. A middle-aged home builder, sports enthusiast, and karate student, my friend found himself in the role of research writer. The point of this story is simply that research writing is a fact of the modern world. Nearly every profession requires this kind of writing: medicine, law, music, art, business, journalism, science, politics, education. You can expect to find yourself both reading and writing "researched" reports in almost any field, regardless of your official job description. Not only your job, but also your hobbies and interests may involve some kind of writing that will require you to collect, organize, synthesize, and even publish documented information. As a student you will find research writing a major method of communicating with your instructors, as well as one of your most important means of learning. Nothing teaches quite so well as the research project you undertake yourself. Once you develop good research skills, you will be an independent learner; you will have joined the community of scholars, researchers, and writers.

Writers of every kind rely on research to provide the facts and figures with which they work. Research writing isn't limited to term papers. Almost all nonfiction writing today relies on some kind of research. Your daily newspaper is a triumph of modern information gathering and reporting techniques. Free-lance nonfiction writers like Tom Wolfe and Tracy Kidder spend months gathering information on America's space program and the development of the computer industry, and then they write books like *The Right Stuff* and *The Soul of a New Machine*. An astronomer like Carl Sagan gathers notes on the evolution of the human brain and then writes *The Dragons of Eden*. Biographers (sometimes even autobiographers) collect memorabilia, documents, and memories, then fashion them into a finished work. A great deal of business writing, too, requires facts, data, information: proposals, feasibility studies, accountability studies, projections, progress reports.

The common thread in all this isn't the end product—not all research leads to a term paper any more than all research leads to a scientific report. What all research writing has in common is the process of research: the collection, organization, synthesis, and publication of information. (I use "publication" to mean presenting to others, not necessarily with the aid of a commercial publisher.) Research writers are part scientist, because they believe it's important to know the truth; they are part adventurer, because they believe that the discovery of truth is an unpredictable adventure, full of interesting side trips and unexpected developments. There is a certain amount of dreary routine in any work, but on the whole, research writing is a rewarding process of discovering the unknown and unanticipated. It's a truism that writers must know their subjects well. You must become knowledgeable about the subjects you write about; you must acquire a large collection of information. Much writing, therefore, is research writing—writing with and about research. Most writers are simply working people who have learned how to inform themselves, how to teach themselves what they need to know in order to write. You may know nothing at all about chimpanzees, but with enough research you can become an expert on these animals. All the world's knowledge on chimpanzees is available to you through research. Nonfiction writers don't invent information: they collect it.

As a student you may not be ready for large research projects, but you can teach yourself the methods and habits of a serious researcher. The chief virtue of research writing is its accuracy, and students are capable of accurate work. All researchers have had to learn to take the time to cross each *t* and dot every *i*. No chemist would say that a drop or two more or less of a chemical won't make a difference. Chemists learn to write reports with as much care and accuracy as they use with measurements in the lab. It's the habit of careful work, the tradition of precise language, that identifies the serious researcher.

The Research Writer's Journal

Like other writers, research writers need practice in writing; they need a place to record thoughts and ideas; and they need to be able to take notes on their readings and observations. The journal is an excellent place for all three activities.

Writing isn't something most people can do on an occasional basis. You must have a regular routine for writing. Professional writers say they write each day at the same time and place, just to keep the juices flowing. You must remind yourself that you aren't a machine: you don't simply crank out facts and figures whenever a writing situation comes up. Instead you are a living, thinking writer. Research writers have as much concern for the strength and flow of their sentences as any other writer. The substance and logic of paragraphs concerns research writers no less than other writers. The tone, design, and conceptual depth of a piece of writing mean as much to researchers as to anyone else. Researchers are writers.

You need a place where you can keep a record of your thoughts. More important, you need to develop the habit of thinking through writing, thinking *while* writing, so that thinking and writing become synonymous. A journal is a good place to talk to yourself (sort of thinking out loud), ask and answer questions about your research, make notes. Your journal should reflect your mind, a catchall for all those fleeting impressions, images, and half-formed thoughts that flash through your cranium day and night. Most people think they have nothing to write about because they make no use of the material they do have. A half-formed thought isn't much, but it's better than no thought at all. If you capture your thoughts on paper, your chances will improve for developing such thoughts more fully at a later date. One thought leads to another, and soon you are off on the trail of something interesting. But only if you have marked the trail, in writing, will it be of much use to you later. It's pleasant to sit and daydream, but a few minutes later the daydream will have vaporized, leaving no trace of your thoughts. Thinking on paper is an important habit to develop.

Finally, writers who read need a place to react to their reading. You can take formal notes, make outlines, jot questions to yourself about the books and articles you are reading, but you also need less formal, more creative kinds of reactions to reading. Think of reading as a dialogue between you and the authors. Like an intelligent participant in a dialogue, you need to respond, agree or disagree, suggest additional examples, point out contradictions or exaggerations, pursue side issues and digressions. One kind of reading suggests the author is using a code to send you a message, and your duty is to receive the message as accurately as possible. Another kind, however, suggests that the author is striking off sparks with a hammer on an anvil, and you must try to grab whatever spark you can. In the one case you permit the author to control you; as a passive receptacle you just take it all in. But in the other case, you are in control, taking whatever ideas appeal to you, running off in whatever direction your thoughts take you. It all depends on what you are reading, your purpose in reading. Your journal can be simple notes or it can be a more creative interaction between you and your reading. (See Chapter 3, Reading Research.) Read the following excerpt from a reading journal. What benefit is this journal entry to the student?

READING JOURNAL

Fouts, Roger S., Fouts, Deborah H., and Donna Schoenfeld. ''Sign Language Conversational Interaction Between Chimpanzees.'' Sign Language Studies 42 (1984): 1-12.

In this study, Fouts et al. report that chimpanzees use signs to each other spontaneously. The chimps were Washoe, Loulis, Dar, Tatu, and Moja (2). The chimps sign to themselves (talking to themselves?), to each other, to pictures, to passing persons and objects outside their enclosure, and to their trainers—who speak English at all times (4). The trainers used only 7 signs: what, want, where, who, which, sign, and name (5). (It was a kind of control, avoiding cuing the apes, use only a minimum number of signs.)

''Because all of the conversations we observed [between chimps] included some nonverbal behavior it may not be possible to define a face-to-face conversation that is so purely verbal that it is totally free of nonverbal behavior'' (5). Not sure what this means, except maybe it is hard to interpret ''signs''—can't always tell if chimp is ''signing'' or scratching?

I have a comment somewhere about ape researchers' inclusion of natural gestures as ''signs''—apes sometimes credited for more signs, more language than they have, really, when natural gestures like waving arms are counted as signs.

Part of this research was concerned about whether apes would talk to each other with their new language. Seems like they didn't: ''Very few conversations with a sign or signs occurring in each turn were observed'' (5).

Other researchers say that apes don't return signs to each other = meaning of ''conversation'' is unusual. Fouts uses example: one chimp approaches

another, a sign is given and apparently ignored. Fouts records this ''conversation''—
1 approach; 2 scream; 3 ''SMILE''; 4 turn away (?)
He says, ''The vast majority of conversations including signs were a mixture of signs and nonverbal actions like this'' (6).

Fouts reports most ''conversations'' were the ''two-turn'' type = one ape initiates with a sign or gesture (takes a turn) and the addressed ape ''takes a turn.'' There are very few ''extended'' dialogues. Fouts gives one example.

Washoe (to Dar): HUG
DAR: HUG (approaching Washoe)
Washoe: HUG
Dar: (approaches, and sits next to her): HUG HUG
Washoe: HUG, COME HUG
Dar and Washoe embrace (10)

Other researchers have commented on inclusion of ''hug'' as a sign = it is a natural gesture among wild apes. (Find source.) So, is this a conversation? Are Washoe and Dar using signs with each other?

''It is interesting to note that Loulis was the primary initiator of signing conversations and that he is the only chimpanzee who has acquired his signs from chimpanzees and not from human companions'' (11). The inference isn't clear—Loulis was also the youngest ape, Washoe's adopted child; he was the most dependent. I don't get what point he is making. Not sure whether this study means anything. If one of the tests of language is whether the apes will use it to talk to each other, it looks like this says no. Or if apes have even two word conversations, is that important?

Darrin Fischer

ACTIVITY 1 Begin your journal. You might begin by listing the subjects that interest you—possible research ideas—or you might list all you know on a particular subject. Write every day. Use your journal to take notes on your research topic: to gather a bibliography; to make outlines, rough drafts. (Give a page reference for every note from a publication.) You should start using your journal to discuss your reading, think about the ideas in what you are reading, ask questions, make observations.

THE REQUIREMENTS OF RESEARCH WRITING

Research writing can refer to a paper as long and formal as a history of legal precedents, or it can mean a paper as short and informal as a documented essay. Research writing covers a wide range of types of writing, levels of formality, and uses of documentation. A short newspaper item, for example, can seem very informal and hardly deserving of the term "research."

> After five years of research, a Columbia University psychology professor has concluded that apes can't talk. More accurately, Herbert S. Terrace found that chimpanzees, gorillas, and orangutans can't seem to get the hang of constructing a sentence.
>
> Such news came as a shock to fans of Washoe, the chimp who seemingly mastered sign language, and Lana, who had learned how to use a computer. Terrace concluded that they weren't learning language at all. All they were really doing, he said, was mocking their trainers.
>
> "There is virtually no evidence," he said, "that domestic animals can produce sequences of the words they have been taught . . . in any systematic manner." Their long sentences told no more than their short sentences, he said.
>
> Our sympathy is with the chimps. Anyone with experience reading government documents (or economics textbooks) knows that long sentences may be totally meaningless.
>
> It seems possible that chimps might be telling us they have no interest in what humans are teaching—and that if learning language got us into the messes we are in, they don't want to follow suit. After all, who wouldn't rather worry about bananas instead of inflation and atomic submarines?
>
> "Apes: Scientists Say They Can't Talk, But It Depends on How You Look at It." *Detroit Free Press*, 3 Nov. 1979: 8a.

This short newspaper article may not seem very scientific, yet the writer had to *find* the information, *organize* it according to journalistic principles, *synthesize* it (assimilate the material and express it in his or her own words), and *publish* it. Research can be written in the first person, the second person, or the third. It can be very formal and abstract, scientific or folksy in tone. It can cover everything from a writer's private journal to a theoretical discussion of technical concepts. Some researchers are very scrupulous about identifying the source of every word, fact, and idea they have borrowed; others are very casual about substantiation and documentation.

ACTIVITY 2 Describe the "Apes" article above. In what ways can it be considered research? In what ways is it different from more formal, traditional kinds of research? Paraphrase or quote from the article to support your points.

The controlling decisions in research writing are based on the writer's situation. (See p. 9.) You must determine who your audience is, what your purpose is. The more scientific or academic your audience and purpose become, the more your readers will expect a formal tone, attention to the rules of evidence, and careful use of documentation. The more objective and information-dominated the writing situation becomes, the more your readers will wish to be able to check the accuracy of your facts and figures and the reliability of the judgments and interpretations in your writing. Research writing is verifiable writing, and the more academic the writing situation is, the more we wish to verify the writing.

Research Must Be Carefully Designed

You must plan your project carefully. You must find a precise research question, and you must create a plan for answering your question. Often researchers must write a proposal so that employers, funding agencies, or clients can see how well the project has been designed. Students, even more than professional researchers, must plan their research projects very carefully to avoid the disappointing results of lackadaisical approaches to writing.

Research Must Be Carefully Implemented

The researcher must have patience and stamina. You must work carefully enough to be thorough. Research can't be rushed, and researchers must learn to live with the frustration of slow and painstaking plodding through research materials. Use systematic procedures in collecting data. Learn to use the research tools available to you in your library.

Research Must Be Carefully Reported

The design and implementation stages must aim at the reporting stage. Writing the paper isn't merely the last stage: it's the culminating stage. The final paper must be a consideration throughout a research project, and the paper itself must be carefully written. Writing the paper is such an important aspect of the research that it sometimes seems larger than the research itself. No matter how good the research is, it can all be lost in a poorly written paper.

Unread research has little value; unpublished research has no value at all. Modern research is the result of relatively inexpensive publishing and a general attitude that all research is worth publishing. We now have a huge community of researchers who are free to work on small problems and free to publish the results of small studies. The availability of research means that anyone can know what other researchers have already done or are presently doing. Researchers can contribute their work to lines of research already established without accidentally duplicating each other. The key to modern research is the assumption that researchers will seek out and read the work that precedes them. Thus many scientists take the position that unpublished or unpublishable research isn't research at all. Badly written papers obscure and confuse research and are generally thought to be worthless.

While in school, many students discover that research writing can be rewarding on its own and can lead to a good deal of satisfaction. Research is democratic in the sense that it allows anyone, regardless of training or social position, access to knowledge. You can become increasingly competent at investigating serious issues, and you can find significant information that others will respect.

Substantiation: The Basic Requirement

The basic requirement of all research writing is the principle of substantiation. We don't count personal opinion or mere assertions as factual statements. Factual statements are those we can verify, and substantiation is the principle of supplying the means of verification. It isn't a fact that "The chimpanzee Washoe can name objects presented to her." But it is a fact that "According to the Gardners, Washoe can name objects presented to her" (38). Substantiation means offering data we can verify, supplying sources we can check. In this case we can't check whether Washoe can actually name anything, but we can check whether the Gardners ever *said* such a thing. The page reference allows us to find the source of this data. Substantiation constitutes a kind of proof. Most often the proof can be stated as Who said so, or How do you know? Thus, to supply substantiation most often means telling your reader where you got your information. Substantiation can be contrasted with documentation, which refers to the use of bibliographic citations. Technically the two things are different: substantiation is the proof being offered; documentation is where you found your information. However, in practice the two are often treated as the same thing.

On the one hand, student researchers are well advised to provide an abundance of substantiation. On the other hand, there is a good deal of flexibility in this principle where professional researchers are concerned. The use of substantiation is very much dependent upon the writing situation. Popularizations written for the general public tend to take substantiation for granted and make much less use of documentation. Facts in the newspaper, for example, may or may not be substantiated; they are rarely documented. But when researchers are writing for the academic community, trying to establish the facts in a research issue, both substantiation and documentation are taken very seriously. Between the two extremes—writing for the general public and writing for the academic community—there is greater or lesser use of substantiation and documentation.

ACTIVITY 3 Read the following excerpt from a student research paper. Where has the author provided substantiation? Where do you see information that requires documentation? (The paper has been altered for the purposes of this activity.)

> There are major social problems inherent in the computer revolution, even if all the technical and intellectual problems can be solved. The most obvious is unemployment, since the basic purpose of commercial computerization is to get more work done by fewer people. For many of the same reasons that the light bulb was feared one hundred years ago, many people have decided that the computer is an object to be loathed. "If you can believe the most alarming of the technological and sociological predictions, a coming generation of 'electronic brains' . . . will leave its feeble creators far behind, reducing mankind to pitiable robots scurrying at the beck of mechanical masters and digital demons." There have been a number of studies done to illustrate this point. One British study predicts that "automation-induced unemployment" in Western Europe could reach sixteen percent in the next decade. "While many high-tech occupations have rapid growth rates, they will account for only seven percent of the new jobs created between 1978 and 1990." It appears that computers will have an impact on the unemployment figures in the future as well as now.
>
> Excerpt from "Computers in Society," Debra A. Usteski

CYCLES IN THE RESEARCH PROCESS

Research doesn't often lend itself to neat, predictable methods; most researchers develop their own procedures and techniques. Those who already know their subjects well may take shortcuts, but students, at least at the outset, will find the research process less confusing if they go through each cycle carefully. I use the word "cycle" here because the research process is seldom linear. It's seldom possible to proceed in a step-by-step fashion so that once you have done step one you can go on to step two and need not worry about step one thereafter. Most researchers report that the cycles are all operating at the same time and that there is a good deal of moving back and forth between cycles. Eventually you will complete each cycle, but it isn't easy to predict when. You may find yourself rethinking early cycles right up to the end of the project. This chapter, for example, was the first one I worked on, as well as the last one, and I have returned to it several times throughout this project.

Cycle One: Analyzing the Research Situation

All writing situations require you to analyze your subject matter, your audience, your view of yourself, and your purpose in writing. Research writing is no different.

Understand Your Task

If your research task has been assigned by someone else, like an employer or instructor, you must first make sure you understand the assignment. Often those who assign research tasks assume you know standard terminology and conventional procedures. Be sure you understand what is required of you. A report is usually an objective compilation of data, but people tend to use the word rather loosely to mean many different things.

Understand Your Subject

Much depends upon your analysis of the subject. In college, the general subject area may be assigned: one of Shakespeare's plays, an aspect of collective bargaining, computers in science, or some other subject related to college work. This subject is only the general area and not a specific research question or topic. Sometimes the subject is totally unspecified: a research paper on any subject related to course work. In that case you must search through the course materials and your own interests to come up with a topic.

Researchers seldom "think up" research subjects. Unlike essay writing, in which we expect writers to be creative and rely mostly on their own ideas and information, research writing is heavily dependent on previous research. Look for research subjects in the library; use the card catalog and the periodical indexes to find books and articles on your subject. Encyclopedias are also excellent sources of information, especially the encyclopedia yearbooks. Research librarians are helpful and can show you how to find material related to your subject. You don't usually have to invent research questions; you *find* them, and the place to find them is the library. If you have complete freedom of choice, you should use the opportunity to research something that holds personal interest for you.

ACTIVITY 4 Make an interest inventory. List as many topics as you can think of that interest you, topics you might care to research (at least a dozen). If you can't think of any, go to the library, thumb through a recent volume of the *Readers' Guide*; select topics that look interesting to you. (See Figure 2-9 in Chapter 2.)

Understand Your Audience

You must consider who your audience is. Topics that interest you personally may not interest your audience, and while you may be satisfied with a summary overview of the subject, you may be sure some audiences will not. The more expert your audience is, the more they will expect you to follow the rules and conventions of research writing. It's wise to find out as much as you can about such an audience. Often professors have very definite expectations, which you can only find out by asking. In some cases you may find yourself writing to a "popular" audience, an audience of other students or any audience composed of nonexperts on your topic. You must consider how to present your material most effectively for such an audience.

ACTIVITY 5 Popular writing versus academic. Read the two excerpts below, and list the differences between them. Be prepared to discuss your list. Which differences can be attributed to the authors' views of their audiences?

> POPULAR WRITING My father never had an office nurse or a secretary. The doorbell was answered by my mother or by whatever child was near at hand, or by my father if he was not involved with a patient. The office hours were one to two in the afternoon and seven to eight in the evening. I remember those numbers the way I remember old songs, from hearing my mother answering the telephone and, over and over again, repeating those hours to the callers: there was a comforting cadence in her voice, and it sounded like a song—one to two in the after*noon*, seven to eight in the *even*ing.
>
> Lewis Thomas, *The Youngest Science: Notes of a Medicine-Watcher*

> ACADEMIC WRITING In the second edition of *A Return to Vision*, we have attempted to correlate the interrelated thematic approach of the first edition more specifically with the requirements of teaching composition. We have done this primarily through the inclusion of a variety of aids to enable the student to understand more clearly the relationship between the concepts and ideas of a given work and the stylistic techniques and structural principles by which they are conveyed. We have also made several significant changes in the table of contents, including the addition of seven new essays and the deletion of a number of selections which our experience and that of other instructors who have used the first edition have shown to be difficult to treat within the framework of a composition class.
>
> "Preface to the Second Edition"

Understand Your Voice

Some research writers adopt a distant, objective voice, regardless of their subject or audience. You can call such voices academic, although some academic writers protest that such writing is simply objective (or "stuffy," depending on your point of view). With the objective voice researchers avoid the pronoun "I" and attempt to present information as if no human agent were involved. The voices writers use in research writing range from very personal and subjective to very formal and distant. Which voice you use depends on your writing situation, of course, but for most research writing some middle-level voice is best: not too subjective, not too distant.

A PERSONAL, SUBJECTIVE VOICE

Before me the creek is seventeen feet wide, splashing over random sandstone outcroppings and scattered rocks. I'm lucky; the creek is loud here, because of the rocks, and wild. In the low water of summer and fall I can cross to the opposite bank by leaping from stone to stone. Upstream is a wall of light split into planks by smooth sandstone ledges that cross the creek evenly, like steps. Downstream the live water before me stills, dies suddenly as if extinguished, and vanishes around a bend shaded summer and winter by overarching tulips, locusts, and Osage orange.

Annie Dillard, *Pilgrim at Tinker Creek*

A MIDDLE-RANGE VOICE

This is the giant ant-eater. Its eyesight is very poor, its hearing scarcely more acute, but its sense of smell is excellent and it is able to locate termites by the scent of their dried saliva mixed in the walls of their mounds. Once the nest is discovered, the ant-eater widens the entrance of one of the main tunnels with the long curved claw on its foreleg and inserts its snout. From the end of this comes a long thong-like tongue which whisks down the termite corridors at great speed, sometimes as frequently as 160 times a minute. Each time it flicks out, it carries a fresh coat of saliva, and each time it is withdrawn it brings with it a load of termites.

David Attenborough, *The Living Planet*

AN IMPERSONAL, ABSTRACT VOICE

The theoretical approach of generating density functions of random variables, discussed above, becomes much more complicated if each Xj in Equation (1) has a distinctly different density function. But even the simpler cases already discussed are usually beyond the mathematical capability of most undergraduate students. Fortunately the wide availability and utilization of computers today makes the derivation of complicated relationships almost procedural.

Athanasios Vasilopoulos, "Computer-Generated Density Functions for Sums of Independent Random Variables," *Collegiate Microcomputer*, Summer 1985

Many researchers and professional-journal editors today say it's acceptable to write in the first person, and nearly all researchers say you must make every effort to make your writing clear and readable. Research is difficult enough without burdening it with poor writing.

ACTIVITY 6 Describe the differences in the three voices above. What self-attitude or self-image causes these differences? How does each author's voice relate to his or her perception of audience and subject matter?

Cycle Two: Beginning the Preliminary Reading

Once you have a general research subject, you can begin looking for your specific research question. Researchers don't invent specific questions. All research questions reside in the research, and you must find the one that interests you. You should begin reading books, magazines, and newspaper articles before making up your mind about your specific research question. Let the data help you. Researchers must find the truth, not impose their own view of truth on the data. It's a mistake to decide in advance exactly what your research project should be. Don't decide at the outset to "prove that cigarettes cause cancer" or that "the military draft should be reinstated." Go to the library first; do some preliminary reading in the general area that interests you. As you read, you will begin to understand the subject and you will begin to see possibilities for research.

There are several advantages to letting the data help you in this fashion.

- First, you avoid biasing the data (avoid imposing your own slant on the data). Researchers shouldn't start out to "prove" anything: your only task is to find what you can.
- Second, you will learn from the preliminary reading what research has already been done and what is still ongoing. You can avoid duplicating other work. In some cases, other researchers will suggest (in their books and articles) areas of research that remain to be explored.
- Third, you will learn the true *scope* of your subject, and this will help you to make intelligent decisions about your project.

Until you become more knowledgeable about your subject through preliminary reading, it's a good idea to keep an open mind. Let your data lead you to your question.

Cycle Three: Framing Your Research Thesis or Question

On the basis of your preliminary reading, you can begin to shape your question. A thesis has two components: a subject (e.g., labor unions) and a question about it (How have they affected small businesses?). The formal research thesis is an arguable proposition, a statement about which reasonable people could disagree. Even if you are later going to word your thesis as a statement ("Labor unions have adversely affected small businesses"), it's a good idea at the outset to state it as a question: "How have labor unions affected small businesses?" Technically, researchers can't know exactly what the evidence means until they have read all the data. Several preliminary sources may suggest, for example, that small cars are unsafe, but you should wait until you have seen the rest of the data before deciding whether this is a good thesis.

College students are faced with a number of problems that affect research. You have a strict limit on the amount of time you can devote to your project, usually not more than a semester or quarter. Then, too, other course work must be done; you are limited to the research you can do in the time you have. You are also limited in the resources you can use. These will vary from school to school, but in general, students are limited to the materials in the school library. Finally, you are limited by the size of the research project itself: the larger the research question, the longer the paper will have to be to treat it adequately. The size of a research question includes complexity, clarity, and other psychological dimensions involved when we say something is a "big" question. Perhaps "scope" is a better term than "size": The scope of the question is defined by the amount of information required to answer it thoroughly.

Your preliminary reading must help you find not just any interesting question, but the smallest possible question. Even the smallest question is subject to a research phenomenon well known to experienced researchers: namely, as you investigate and explore, the research question will grow larger, more complex, and less easy to answer. What starts out as a simple question that seems almost too obvious and too simple turns out to be not at all what was imagined at first. It's this process of discovery, after all, that makes research interesting and rewarding. Use the four guidelines for the thesis to help you refine your question.

FOUR GUIDELINES FOR THE THESIS

Specific

Researchers may start with vague notions, but soon the research question must become clear and specific. Students should avoid research questions involving abstractions, intangibles, and large concepts that are hard to define. Subjects like *truth, honor, justice, democracy, moralism,* and so on mean different things to different people and will not make suitable topics for limited research. A topic like "How has the decline in Western civilization been related to the decline in the study of the humanities?" sounds very interesting, but is much too broad. General subjects like *modern industry, urban life, women's clothes, drugs,* and so forth deal with groups and categories of things. They should be broken down into smaller components or subcategories. One test for size or scope of subject is to see whether you can name a specific example under the general category: Instead of all of modern industry, for example, select a specific example like the oil industry, food services, automobile manufacturing, and so on. You must try to name an unambiguous subject.

Limited

It's possible to write about large subjects in summary fashion—you could write a short paper on "The effect of foreign trade practices on the American stock market"—but few would call that research.

- For research we require a *small* subject done in great depth. It's always assumed that professional researchers have read everything related to their research, even if that means hundreds of books and articles. For limited research, you need an absolutely limited subject, meaning the smallest one possible.
- If you look through the card catalog or periodical indexes, you will see that each subject heading is subdivided into smaller and smaller divisions. One of these small subdivisions may become your research question.
- You may start out with the general subject of college sports, for example, but you will soon discover there is too much material to deal with all college sports. If you limit yourself to hockey, your subject will be much more specific, but still too large. You will find articles on the history of hockey, the economics of hockey, hockey players, hockey equipment, and on and on.
- You are required to write a thorough paper, but a thorough paper on hockey would be at least the size of a book. You must limit yourself to one small aspect of hockey.

The most limited questions you can ask about anything are usually those that can be answered yes or no. Using your reading about hockey as a guide, pose yes-or-no questions for yourself: Is hockey gaining in popularity? (Yes or no?) Is hockey too violent? (Yes or no?) Are hockey players overpaid? (Yes or no?) You may find that your question can't be answered with a simple yes or no. Many questions turn out to be too complicated for such simple answers; but at the outset, this simple yes/no formula will help you to limit your research question.

Worthwhile

Obviously there are many questions you could ask about any research subject. But are they all equally worthwhile? Should you select one with the flip of a coin? Research can be quite demanding and time consuming; if you are working on a question you aren't completely interested in, you will find it weary work. You must select a question that is worthwhile not only to you, but to others as well. Resources are limited. Skilled researchers often have to justify the money, equipment, and hours their work requires, and college students are even more limited in resources. You need a thesis that is worthwhile in relation to your college course work and worthwhile in relation to the body of research (in the library) from which it comes. From your reading you will discover what has already been done, what lines of inquiry are being developed, and what is or is not considered important in the field you are studying. You will gain some understanding of the relative importance of the subject from the research itself. Beyond that you must consider the purpose and the audience of your research. Unless you are told otherwise, you should assume you are writing to a general audience—the educated general reader—for the purpose of enlightenment. You should avoid overworked subjects, and those not likely to interest educated adults. Trivial, immature, boring subjects should be avoided.

Researchable

If there is little or no material on your subject, it isn't researchable. College students should not plan to borrow much material through interlibrary loan. If your own library doesn't have ample material, you should choose another thesis. Think twice about subjects that are available only in books or only in newspaper articles or in some other limited sources. While your library may have the sources, the fact that they are limited suggests the subject may not have been researched enough for your purposes. The best subjects are those for which you can easily find sufficient material in your own school library.

Some data may be available in very technical or complex language, such as that found in advanced scientific journals or doctoral dissertations, but such difficult material is usually not suitable for students. Unless you can understand your materials easily and thoroughly, the subject should be considered not researchable, despite the availability of materials. That is, the readability of the material is a research limitation for most students—the more difficult the readability, the less researchable. Read the following example; anything written at this level is too difficult for anyone but an expert:

> The fulcrum of our profound phenotypic and adaptive disunion from the chimpanzees and the gorilla must lie in the differential timing of gene expression during brain development. The regulatory changes have a retarding influence upon our unfolding. Fetal growth rates eventuate in hypertrophy of the organ-complex controlling (among other effects) human linguistic competence for speaking and in processing the speech of others, in brief, the language-using animal's species-specific behavior.
>
> Umiker-Sebeok, J., and T. Sebeok. "Clever Hans and Smart Simians: The Self-fulfilling Prophecy and Methodological Pitfalls." *Anthropos* 76 (1981): 89–165.

ACTIVITY 7 Evaluate the following thesis statements and questions. Which might be a good thesis for a limited student paper? Explain any faults or problems you find in any of the statements.

1. Sexual superiority: Is there a genetic code for success?
2. Quadraphonic reproduction: The technology of four-channel transmission

3. Which is the smartest creature in the animal kingdom?
4. Cannibalism: A psychosexual disorder with abnormal physiological concomitants
5. The Salem witch trials: Was justice served?

THE RESEARCH PROPOSAL

Because research is difficult and frequently costly, an essential feature is the research proposal. Even though the research may be requested by someone else, the researcher is usually required to write a proposal. The bigger the research project (such as a project to build a new bomber for the government), the more complex and thorough—and essential—the proposal will be. However, small research projects also benefit from a proposal. If you are writing for an employer, department supervisor, or college instructor, your proposal will give him or her a preview of what you plan to do. It will permit critical feedback at the outset of the project and will help you to head off research problems before they arise. The point of a proposal, or prospectus, is to provide a basis for allowing, disallowing, or changing a research project. The proposal ensures that you have done your preliminary reading and that your project is research-based, not just an idea off the top of your head.

There are no absolute rules about prospectus writing. In general, you must describe what you propose to do, why you want to do it, how you will do it, and what you will do it with. But there are many variations of this general prospectus, depending on who your audience is, how big the project is, and the technical nature of the project itself. The credibility of the researcher is at stake in the proposal: you must convince someone that you are capable of doing the project. Your proposal must look as thorough, accurate, and professionally prepared as you can get it. (See example, "A Research Proposal," pp. 20–21.)

Background

The background to a research project answers the question Why? Why should this research be undertaken? Research projects are seldom random ideas that pop into someone's head; the source of a research project is seldom merely the personal interest of the researcher. A project on chimpanzees, for example, shouldn't be undertaken just because you are personally interested in apes. Preliminary reading about apes will reveal areas that need research, problems or questions about apes that haven't yet been adequately researched. A little reading in the library will soon reveal a background of research in the ape-language question; you can see what has already been tried, mistakes that have been made, suggestions at both practical and theoretical levels. The background of a proposal is usually a history of previous research.

Description

You must describe what you plan to do. If you are going to construct a questionnaire and mail it to scientists working with apes, you must say so, and describe the questionnaire as well as you can for the reader's benefit. If you are going to do a statistical analysis of the answers to the questionnaire, you must describe the statistical tests you will use, and why. Since much research is done in the library, you should state that you intend to find and collect the available research, read it, and then draw conclusions based on your reading. The description of a research project answers the question "What will you do?"

You must pay close attention to the language you use; words like "explore," "discuss," and "analyze" mean different things to different people. "My project will be a study of chimpanzees and language" is too vague. What do you mean by "a study"? See the guidelines for thesis questions, p. 14.

The proposal should always be written as if the project were your own idea. That is, even if your professor has given you the assignment, you must describe the intent of the project as what *you* want to do, not what the professor wants you to do. The proposal shouldn't be written with the assumption that your instructor knows what you are supposed to be doing. Anyone, not just the instructor, should be able to read your proposal and learn from it what your intent is.

Procedure

The Procedure section of a proposal is sometimes an elaboration of the description; it can be written as part of the description. But it's often presented as a separate section, sometimes called Method. If you are going to use a questionnaire, exactly how will you create it? Questionnaire construction is a specialized branch of research, and you must show that you are aware of the techniques and problems in this kind of research. If you are going to use library materials, which ones? How many? How will you find them? What will be your collection technique? It's a good idea to list here the different kinds of materials you will seek: books, journals, government documents, and the like, and how you will find them—and perhaps why, and what indexes you will use. If the majority of your research consists of collecting materials in the library, you must be as detailed as possible about how you plan to collect them. Your reader needs to be reassured that you are familiar with the materials in the library and have a plan for using them. The reader needs to be reassured that your procedure is likely to produce the result you want, that your procedure will lead you to the answer to your research question.

Significance

What is the significance of your research question? Why is it worth answering? What are the advantages of answering it? If you were attempting to get a research grant, for example, you could not present your question as merely "interesting." Nor can you assume there is any such thing as self-evident significance. It may be clear to you that your project is significant, but it's the reader you must convince.

One answer is that it's better to have facts and figures than belief. If the research merely confirms what we already believe, we will at least know why we believe as we do. Perhaps the research will uncover some specific problems with apes that we might solve; if apes resist learning language in a lab setting, maybe we could change the setting or improve it. But why should we care? Why should we try to teach an ape to talk? If we can teach an ape to talk, maybe we could teach retarded or injured humans with the same technique? Perhaps from this kind of research we may learn something about human language or intelligence or culture? The significance of a research question, like the question itself, must be found in the research. Significance is not your opinion; what do the researchers say about it?

Problems

It's important not to obscure research problems. If your project will require money, the purchase of expensive equipment, the use of unstable chemicals, or some other troublesome problem, you must say so. The general rule is that you mustn't hide or diminish any problem that would cause your instructor to disallow your project if the truth were known. It's best to state the problems clearly and honestly and explain how you will deal with them.

If you are doing library research, the only real problem is the possible inaccessibility of materials. But this isn't an inherent problem; with proper planning or a little resourcefulness you should be able to work around this difficulty. Therefore, library research projects don't usually have a Problems section.

Requirements

If your project has any special requirements—equipment, money, travel, extra time—you must list them in the proposal. If you feel it will be necessary to travel to distant places for interviews or for material not available locally, say so in the prospectus. There is only one place on this planet where you can study the giant panda in the wild: you will have to go to China. Library projects usually don't have a Requirements section, but could have if, for example, you plan to make much use of interlibrary loan (not a good idea).

Works Cited

A key component of the proposal is your references. You must document any source references in your proposal, of course; the function of your bibliography is to make sure you have done some preliminary reading and have investigated the availability of information. The research proposal should look like the preliminary work of a researcher who already knows a good deal about this question. Then too, your Works Cited section lets others see how much you have found and whether you seem to be heading in the right direction.

A RESEARCH PROPOSAL

Can Apes Talk?

The Background to the Question

The earliest efforts to teach apes to talk began in the 1930s (McLaughlin). A chimpanzee raised as if it were a human infant learned to recognize about 100 words, but couldn't speak. In the '40s another chimpanzee learned to recognize many words, but could say only ''mama,'' ''papa,'' and ''cup'' (McLaughlin). Serious scientific study on this question began with Professors David and Ann Premack in 1965.

The Premacks taught a chimpanzee to use a ''language'' made of arbitrary plastic symbols. More recently, Allen and Beatrice Gardner taught a chimpanzee to use ASL, the gestural language of the deaf. Chief critic of all this research has been Herbert S. Terrace. Several of his books and articles suggest that the apes are only performing for their trainers, like chimpanzees in a circus.

Description

Since the question: Can Apes Talk? is controversial, my research will investigate both sides of the issue. An important part of this research will be an analysis of what different researchers mean by ''talk.'' It's obvious that apes aren't biologically equipped for speech, and therefore the definition of ''language'' is significant in the research.

The Significance

The most useful result of this research has been the use of Premack's plastic-symbol language to teach retarded children and aphasic adults. Beyond that, Premack is interested in the nature of intelligence itself, and the work with chimpanzees and nonverbal behavior is opening new approaches to these questions: What is intelligence? How does the mind work? In general, we don't know much about how human beings learn to speak (Katz). Research with apes is beginning to shed some light on human language acquisition.

Works Cited

Gardner, R. A., and B. T. Gardner. ''Early Signs of Language in Child and Chimpanzee.'' Science 187 (1975): 752-53.

Katz, Jerold J. The Philosophy of Language. New York: Harper, 1966.

McLaughlin, Patricia. ''Learning from Some Chimps.'' Pennsylvania Gazette, Feb. 1976: 16-22.

Premack, A. J., and D. Premack. ''Putting a Face Together.'' Science 188 (1975): 228-36.

Terrace, H. S. ''How Nim Chimpsky Changed My Mind.'' Psychology Today June 1979: 65-76.

ACTIVITY 8 Do some preliminary reading to find a research topic that interests you; read half a dozen or so articles on your topic. Write a proposal for a semester-long research project. Write your proposal objectively, but without hedging. Avoid vague qualifiers like "hopefully," "possibly," "maybe," and the like.

Cycle Four: Start Collecting Your Data

Begin building your bibliography, reading, and taking notes. As you read, you may find that you need to make adjustments in your research question, based on the available data. Your original idea may have been to write about "violence in hockey," but as you read you will discover that most of the available information concerns "illegal hits in hockey." You should adjust your thesis accordingly; always be willing to go where the data lead you. You should be in control of the project, of course, but part of your control is an openness to the research, a willingness to give up old ideas for newer ones. You can be systematic about data collection: use a standard library procedure and note-taking method. You must be thorough within the limits of your time: read as much as possible. Ideally, researchers find and read everything, even hundreds of books and articles, on their subject. The more sources and notes you have, the more confident you and your readers can be of your results. A dozen experts who say cigarettes don't cause cancer may seem like an impressive number; however, if there are a hundred others you could have found who assert the opposite, a dozen seems less impressive. (Evidence is discussed in Chapter Four.)

Cycle Five: Finish Data Collection; Write the Paper

Between the data collection and the report writing—or ongoing with both of them—is the analysis of the data. Research isn't so simple that you can sit down with your note cards and begin to type up your report. First you must sort and group the data into an arrangement that means something. The data are bits and pieces of information. Individually they don't mean anything, but through the process of analysis and interpretation of data, the researcher transforms information into evidence. The researcher must weigh the evidence, reason about the data, and reach a conclusion. The real skill of the researcher lies in deciding what the data show; interpreting the data is the most difficult and the most interesting part of research. Interpretation is a matter of judgments; there are some principles and guidelines regarding evidence, but there are no mechanical rules you can apply.

The report, then, is written to show the reader what the data mean. In most modern research that means showing the reader what the data *probably* mean. Given these data, we can only draw "fair" or "reasonable" conclusions. Thus, the paper itself must be written carefully and persuasively to convince the reader that the conclusions are probable.

Because of the cyclical nature of writing, you can't really wait until all the data are in before thinking about your report. As you collect your data, you will begin to see patterns of organization, and you can begin making trial outlines. As you continue to collect data, you can adjust and revise your outlines accordingly.

As you write your first draft, you may discover holes in your outline—areas where you need more information—or you may see something while drafting that you didn't see while outlining. Thus drafting, too, becomes cyclical as you move between drafting and outlining, adjusting both until you finally produce the finished paper, in which draft and outline agree.

When you have a draft that satisfies your research requirements—a fully documented paper that answers your question, establishes your thesis, and so on—you can begin to give greater consideration to your writing. You must read with a hypercritical eye, looking for three things: accuracy, clarity, and economy (or the lack thereof). (See Chapter Seven, Style.) Researchers must revise until their report says exactly what it must say in the way it must be said. Frequently researchers will ask a colleague or friend to review for them, and published research is usually sent to manuscript reviewers, experts in the field, who will give the author critical feedback. Finally, after much revising and reworking, the report will be finished. The final step in the process is proofreading for spelling, punctuation, grammar, and so on. As irritating as such matters can be, mistakes at this level can have a distorting effect on research; they can bias your reader against your research.

STUDY GUIDE: Chapter One The Research Process

1. What is the research process?
2. How might research writing be of interest to students?
3. What is the chief virtue of research writing?
4. What is the function of the research writer's journal?
5. What are the three requirements of research?
6. What is substantiation?
7. Why is "cycles" a better description of the research process than "steps"?
8. What are the cycles in the research process?
9. What is the difference between a general subject area and a specific research question?
10. What use might a researcher make of encyclopedias?
11. What is "voice" in research writing?
12. Why is it a mistake to decide in advance exactly what your research project should be?

13. What are the advantages to letting the data lead you to your question?
14. What is a thesis?
15. How can you determine the size of a research question?
16. How big should your research question be?
17. What is the "growth phenomenon" of research?
18. What are four guidelines for the thesis of a research paper?
19. What are abstractions and intangibles?
20. How do you know when you have an unambiguous subject?
21. How can you arrive at a most-limited question for research?
22. How do you know whether a research subject is worthwhile?
23. How can you tell whether a subject is researchable?
24. In what way is readability a research limitation?
25. What is the purpose of a research proposal?
26. What is the Background section in a research proposal?
27. What are the Description and Procedure sections of a research proposal?
28. What is the Significance section of a research proposal?
29. What are the Problems and Requirements sections of a research proposal? What problems or requirements are likely in library research?
30. What is in the Works Cited section of a research proposal?
31. How much is enough research?
32. What are data?
33. How can you transform information into evidence?
34. What is the most difficult part of research?
35. What are reasonable conclusions?

CHAPTER TWO

Using the Library

Research starts in the library. Before you decide on a research project, go to the library. Let your preliminary reading tell you what your research question should be. What your project will be depends on what it *can* be, and that is dependent on the kinds of information available. New research relies on old; your project must rely on the work of researchers who have preceded you. So before you attempt to write a proposal, before you make up your mind about what you want to do, go to the library. Start your preliminary reading.

We can't know what any given study of ape language means until we can see the whole picture. At the outset, the new researcher is somewhat lost in a great maze of random details; what is missing is context—the overall background that will allow us to evaluate where each new bit of evidence belongs; the relative importance of each book, article, fact, and so on. Thus, it's helpful to find the background research first. Once you have some idea of the general dimensions of your research problem and some of the history of your research question, each new book or article you read will make more sense, and will help you fill in the holes in your knowledge.

THE LIBRARY SEARCH STRATEGY

Researchers, perhaps more than most people, need to be systematic in what they do. You need to develop a systematic search strategy for your library work. Developing a careful routine for the library will ensure that you are both thorough and efficient when looking for data. Without a strategy, you will be faced with an endless stream of apparently random data. A good search strategy will immediately impose the beginnings of order on your research. Researchers are highly individualistic in the ways they do things; eventually you too will work out your own procedures for research, but at the outset you will find it most useful to proceed from general to specific sources.

STRATEGY ONE: LOOK FOR BACKGROUND MATERIAL

Note that the search doesn't usually start with the card catalog. It's too soon to worry about books. A student interested in the question of apes and language needs to know some of the background before getting into serious research. Then, too, there are so many books in the library that you need to know something about your subject in order to evaluate, to make informed selections among books.

The General Reference Works

The library holds a number of reference works under the broad heading of encyclopedias and dictionaries. (See Evidence in Chapter Four for guidelines on use of general reference works.)

The General Encyclopedias

The general encyclopedias are good sources for background and history. Many encyclopedia articles provide bibliographies; you can find out quickly some of the important works and standard references. As a general rule, you should read what is in the encyclopedias first. Other researchers will assume you have read the encyclopedia material; such information constitutes the most general background—it's what is known on most subjects—and you should read as much of it as possible. How you should use this material is another question; see Evidence, Chapter Four.

Always use the encyclopedia indexes. It's pointless to try to guess how anything might be listed. For example, in the *Americana*, the ape-language subject is listed under "Ape," but in the *Britannica*, it is indexed under "Animal communication."

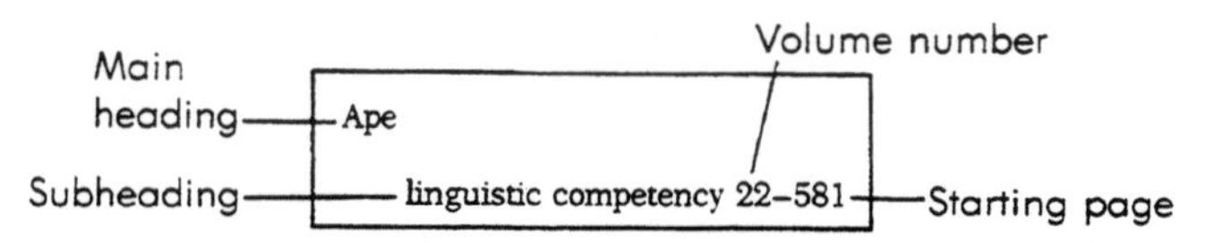

FIGURE 2.1 Excerpt from *Encyclopedia Americana, Index*, 1986.

The index shows that under Ape there is a reference to linguistic competency in Volume 22, starting on page 581. An excerpt from the PRIMATE article discusses relevant information:

PRIMATE
Language. Abstract symbolic naming, conceptual categorization, spontaneous labelling, and meaningful "word" combinations are all elementary aspects of linguistic competency that have been demonstrated by apes through instruction in sign language or synthetic languages. Once a basic vocabulary and stock phrases are learned, the apes are able to use these learned elements in novel ways to make requests of their environment and even to describe it. Lana, a chimpanzee at the Yerkes Primate center, could request, among other things, that food be given directly to her, placed in a bowl, placed in a cup, or placed in a vending device, depending on what was available or on what she desired. She could specify the particular item or activity . . . .

from *Encyclopedia Americana*, 1986

The articles in good encyclopedias today are written by experts, and one easy way to find out who some of the experts are in any field is to find out who wrote the encyclopedia articles. The *Americana* article was written by Duane M. Rumbaugh and Sue Savage (the Lana researchers). Furthermore, many encyclopedia articles include short bibliographies, so you can begin collecting your preliminary bibliography while you are still reading general reference works. At the end of the *Americana* article you will find references to books like these: Geoffrey H. Bourne, *The Primate Odyssey*, 1974; Alison Jolly, *The Evolution of Primate Behavior*, 1972; Adolph H. Schultz, *The Life of Primates*, 1972.

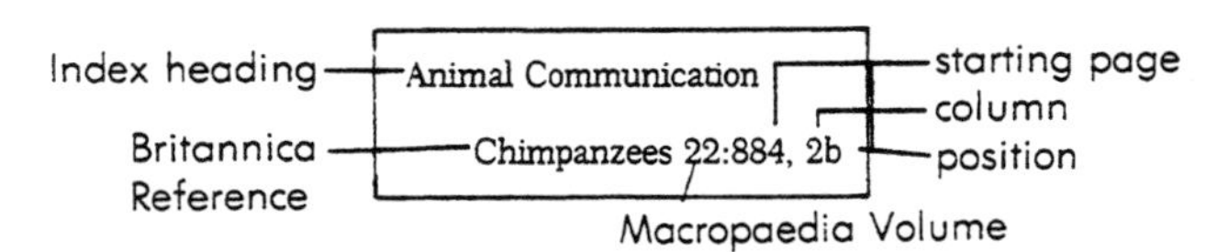

FIGURE 2.2 Excerpt from *Encyclopaedia Britannica, Index* 1986.

The *Britannica Index* shows that under Animal Communication there is a reference to Chimpanzees in Volume 22 of the *Macropaedia* (Knowledge in Depth), starting on page 884, column 2, the lower half of the page (b). The following is an excerpt from the *Britannica* article:

> A number of efforts to teach chimpanzees to communicate information by the use of words have been complete failures. On the other hand, success has been achieved by using a gesture language that is widely used by deaf people to communicate with each other. Two investigators made spectacular progress in understanding the communication of chimpanzees after five years of training a young female named Washoe to use many elements of the gesture language. A great advantage of this technique is that the animal's achievements can be compared directly with those of normal children in English and with those of deaf children of the same ages in learning this particular sign language.

> After four years of experiment, Washoe had learned to use correctly more than 40 different signs for nouns (e.g., bird, clothes, hammer). In addition there were four signs of appeal (hurry, please, etc.), five for location (in, up, down, etc.), and ten for attributes (red, funny, sorry, etc.). Particularly interesting was her ability to use the pronoun "you" appropriately—that is, for any companion—and in combination with a wide variety of signs for actions and attributes. She began to use pronouns and proper names in the third year of teaching and later produced such sign language sentences as "you Roger Washoe out," "you me go out," and "you me go out hurry."
>
> from *Encyclopaedia Britannica*, 1986

The *Britannica* article is signed with the author's initials: W.H.T. Find "Initials of Contributors and Consultants" in the back of *Propaedia*: William Hohman Thorpe, Emeritus Professor of Animal Ethology, University of Cambridge, author of LEARNING, ANIMAL. The bibliography at the end of the *Britannica* article includes R. A. and B. T. Gardner, "Teaching Sign Language to a Chimpanzee," *Science*, 165: 664–672 (1969); D. Premack, "A Functional Analysis of Language," *J Exp Analysis Behav*, 14: 107–125 (1970).

ACTIVITY 9 Select a general subject area that you might enjoy researching. Find several general encyclopedia articles related to this subject. Prepare summary notes on the kind of information you find.

Specialized Encyclopedias

In addition to the general encyclopedias, there are others more limited in coverage.

Encyclopedia of American History
Encyclopedia of Educational Research
Encyclopedia of Religion and Ethics
Encyclopedia of World Art
International Encyclopedia of Chemical Sciences

While these are specialized in the sense that they deal with less general subjects, the information in them constitutes the general background anyone who is interested in these subjects needs to know. Research in the sciences is aimed at other scientists; for a nonscientist, such research can pose problems. A student may know very little about the science that studies talking apes, a branch of psychology. Technical terms, the scientific methodology, and possible problems in the research may require you to look for some background material in psychology itself. Then, too, if you know very little about apes, talking or otherwise, it's probably a good idea to learn something about the animals.

FIGURE 2.3 Excerpt from *Encyclopedia of Psychology*, Vol. 4 *Index*, 1984.

Excerpt from *Encyclopedia of Psychology*, Volume 1:

> ANIMAL COMMUNICATION (D. A. Dewsbury)
> Perhaps the most publicized examples of animal communication in psychology studies are studies of "language" in chimpanzees. Early researchers met with limited success when attempting to condition chimps to vocalize. More recently, however, several researchers have trained animals to use rather complicated communication systems, by relying on gestures and operant responses. Gardner and Gardner (1969) employed American Sign Language, the system used by many deaf humans, and succeeded in establishing a complex signal repertoire. The chimp, Washoe, could both send and receive an impressive catalogue of such signals. Premack and Premack (1972) taught their chimp, Sarah, to position plastic symbols in particular order so as to convey messages. Rumbaugh and Gill (1976) used a computerized system with Lana, who had to depress a set of operant response keys in the appropriate order as part of her language system. The extent to which these systems represent true language, and especially the use of grammatical rules, remains quite controversial (J. L. Marx, 1980 and Terrace, 1979). [Bibliography: Sebeok, T. A. (Ed.) *How Animals Communicate*; Smith, W. J. *The Behavior of Communicating*]

The *Encyclopedia of Psychology* will help anyone interested in this research; it is full of researchers' names and bibliographic references a student could use. Since both the *Britannica* and *Encyclopedia of Psychology* mentioned the Gardners and Premack, it's safe to assume these are important researchers in this area. With a little background research, your bibliography of sources will begin to grow all by itself. And with this much research you may begin to have some idea what your thesis question should be.

Excerpt from *Grzimek's Animal Life Encyclopedia*, Volume 10 *Mammals I*, The Great Apes, pp. 488–502:

> CAN ANTHROPOID APES LEARN TO TALK? Chimpanzees, gorillas, and orang-utans neither have a special inclination to imitate sounds nor do they succeed in imitating the human voice as the parrots or other talking birds are able

> to do. However, while birds are merely able to imitate the sound of the human voice, and are at most able to reproduce the words in the appropriate situation, anthropoid apes are able to use words which they have learned appropriately. The young chimpanzee female "Vicky," who was raised by the Hayes, had learned to say three words; she was able to use two of them correctly, "Mama" for her human foster mother, and "cup" for the object as well as for . . . .
>
> Dietrich Heinemann

ACTIVITY 10 Find one or more specialized encyclopedias related to your research area. Take summary notes on the relevant information.

No matter how simple a research idea seems at the outset, it may soon become more complex. The encyclopedias reveal the background information, the history behind most subjects. Researchers have tried to teach apes to talk and failed. Apes are biologically unable to speak, and that fact is readily available in the encyclopedias. However, the encyclopedias raise a possibility that might not occur to a general reader; perhaps apes can use another kind of language: gestures, computers, plastic symbols? Most research projects grow geometrically; each new thing you read raises new possibilities, new ideas for you to research.

STRATEGY TWO: LOOK FOR BOOKS

After you have read some of the general reference works in your area, you should know enough of the background material to begin collecting books and articles in earnest. You should have a fairly clear idea about your research thesis. It's a good idea to continue reading as much as possible, but while you are reading, you must also continue to build your bibliography. How many items should your bibliography contain? All of them. Researchers try to find *all* the books and articles on their subject. It isn't unusual for a bibliography to contain hundreds of items. Students may not need that many, but it's difficult to know which items are important and which ones are not.

Master Bibliography

The rule on bibliographies is to start with surplusage. Note that this bibliography isn't the same as the bibliography that will accompany your paper; this is your master bibliography. One of its functions is to demonstrate how much material is available on your subject. Its other function is to guide your reading, to make sure you have read everything available, whether or not you use all sources in your paper. Look especially for books and articles that survey the entire field you're interested in. Recent works are likely to review much of the preceding research, so you can get a quick overview of your research area. Look for books and articles that include bibliographies.

Bibliographies

The first step in building your own bibliography is to find the published bibliographies. You should already have found some at the end of encyclopedia articles, but there are indexes to bibliographies, and these will be the quickest sources for you to use. These sources will either list bibliographies for you or tell you where you can find a bibliography on your subject area.

American Scientific Books, 1962–.
An Annotated Bibliography of 20th Century Critical Studies of Women and Literature, 1660–1800, 1977.
Annual Bibliography of English Language and Literature, 1920–.
Bibliographic Index: A Cumulative Bibliography of Bibliographies, 1938–.
Bibliographies in American History, 1960.
Bibliography of American Literature, 1955–.
Books in Print: Author Guide, yearly.
Books in Print: Title Guide, yearly.
Cambridge Bibliography of English Literature, 1941–57.
Chemical Publications, 1965–.
The Critical Index: A Bibliography of Film, 1946–73.
The Cumulative Book Index, 1899–.
Dramatic Criticism Index: A Bibliography of Commentaries on Playwrights from Ibsen to the Avant-Garde, 1972–.
Guide to the Literature of Mathematics, 1958.
A Guide to Reference Materials in Political Science, 1966–.
Harvard List of Books in Psychology, 1971.
Historical Abstracts: Bibliography of the World's Periodical Literature, 1775–, 1955–
International Social Science Institute.
International Bibliography of Economics, 1930–.
International Bibliography of Historical Sciences, 1930–.
International Bibliography of Political Science, 1979–.
International Bibliography of the Social Sciences, 1955.
International Bibliography of Sociology.
MLA International Bibliography, 1921–
Literature and languages.
The Negro in the United States: A Research Guide, 1965.
Paperbound Books in Print
Published twice a year.
Political Science: A Bibliographic Guide to the Literature, 1965–
Supplements, 1966–.
A Reader's Guide to the Social Sciences, 1970.
Scientific, Medical and Technical Books Published in the United States of America, 1958.

Sources of Engineering Information, 1948.
Subject Guide to Books in Print, yearly.
Textbooks in Print, 1926–.
The Women's Rights Movement in The United States, 1848–1970: A Bibliography and Sourcebook, 1972.
A World Bibliography of Bibliographies, 1965–66.

The indexes will quickly provide you with numerous references. For example, see Figure 2.4.

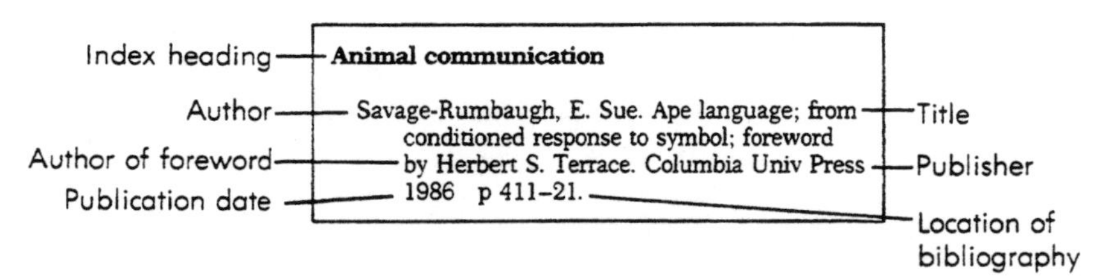

FIGURE 2.4 Excerpt from *Bibliographic Index: A Cumulative Bibliography of Bibliographies*, April 1987

The bibliographic index shows there is a ten-page bibliography in the recent book by Savage-Rumbaugh. Other bibliographic sources provide still more. You should be able to find more than enough sources if your subject is researchable at all.

Library of Congress: Subject Headings

Instead of wandering around the card catalog, which is likely to be immense, you can save yourself a lot of time and frustration by using the *Library of Congress: Subject Headings*. This work shows how the subject descriptors for the Library of Congress (and any library using the LC system) are organized. For example, you won't be able to find the "Kulanapan Indians" by looking under *K* in the card catalog, but the *Subject Headings* will quickly point you in the right direction (*Pomo Indians*). A sample from Volume II (Jury–Z) is shown in Figure 2.5.

Look in the front pages of any index to discover the meanings of symbols and abbreviations. For example, in the *Subject Headings*, *sa* means "see also" and refers either to equivalent or narrower headings, like subdivisions. The symbol *xx* means "see also from" and refers to related (generally broader) headings. The symbol *x* means the term (following the *x*) isn't used; it isn't a heading, but might be mistaken for one. Note the general LC classification numbers after some entries, to help identify branch(es) of knowledge. (They aren't completely reliable, however, due to constant revision.)

Polytheism *(BL217)* *(Continued)*
Monotheism
Pantheism
Religion
Religions
Theism
Polytheism (Islam) *(BP166.22)*
x Shirk (Islam)
xx God (Islam)
Polythene
See Polyethylene
Polytonality
See Tonality
Polytopes *(QA691)*
xx Hyperspace
Topology
Polyunsaturated fatty acids
See Unsaturated fatty acids
Polyurethanes
Polyvinyl chloride
— Tariff
See Tariff on polyvinyl chloride
Polyvinyl plasma
See PVP
Polyvinylpyrrolidone
See PVP
Polywater *(QC145.48.P6)*
sa Liquids
x Anomalous water
Modified water
Superdense water
Water II
xx Liquids
Water
Polyzoa *(Indirect)* *(QL396-9)*
sa Cheilostomata
x Bryozoa
Polyps
Radiata
Zoophyta
xx Animal colonies
Invertebrates
Molluscoidea
Polyzoa, Fossil *(QE798-9)*
sa Cheilostomata, Fossil
Ctenostomata, Fossil
Expletocystida
Gymnolaemata, Fossil
Pomacentridae
sa Abudefduf
Pomacentrus
Pomacentrus *(QL638.P77)*
xx Pomacentridae
Pomacentrus australis *(QL638.P77)*
Pomacentrus pseudochrysopoecilus *(QL638.P77)*
Pomadasidae
See Grunts (Fishes)
Pomadasyidae
See Grunts (Fishes)
Pomar family
x Perez de Pomar family
Pomegranate *(Horticulture, SB379.P6; Pharmacy, RS165.P8)*
Pomegranate in art
xx Art
Pomègues, Île, France
See Pomègues Island, France
Pomègues Island, France
x Île de Pomègues, France
Île Pomègues, France
Pomègues, Île, France
xx Islands—France
Pomelo
See Grapefruit
Pomerania, Eastern, Poland
See Pomerelia, Poland
Pomerania, Polish (1919-1939)
See Pomerelia, Poland
Pomeranian dogs *(SF429.P8)*
xx Toy dogs

Pomerelia, Poland *(DK4600.G44)*
x Danzig-Westpreussen (Reichsgau)
Eastern Pomerania, Poland
Gdańsk Pomerania, Poland
Pomerania, Eastern, Poland
Pomerania, Polish (1919-1939)
Pommerellen, Ger.
Prussia, West (Province)
Royal Prussia
Województwo Pomorskie, Poland
Pomeroy, Iowa
Pomezia region, Italy
Pomme de Terre Lake, Mo.
x Pomme de Terre Reservoir, Mo.
xx Lakes—Missouri
Pomme de Terre Reservoir, Mo.
See Pomme de Terre Lake, Mo.
Pomme de Terre River, Mo.
xx Rivers—Missouri
Pommer *(ML990.P6)*
xx Bassoon
Oboe
Shawm
Woodwind instruments
Pommerellen, Ger.
See Pomerelia, Poland
Pomo Indians *(E99.P65)*
sa Kashaya Indians
Yokayo Indians
x Kulanapan Indians
xx Indians of North America
— Religion and mythology
Pomo language (Eastern) *(PM1601)*
x Kulanapan language (Eastern)
xx Pomo languages
Pomo language (Southeastern) *(PM1601)*
x Kulanapan language (Southeastern)
xx Pomo languages
Pomo languages *(PM1601)*
sa Pomo language (Eastern)
Pomo language (Southeastern)
Proto-Pomo language
x Kulanapan languages
Pomology
See Fruit
Fruit-culture
Pomona Lake, Kan.
x Pomona Reservoir, Kan.
xx Lakes—Kansas
Pomona Reservoir, Kan.
See Pomona Lake, Kan.
Pompadour fish
See Discus (Fish)
Pompano, Florida
See Florida pompano
Pompeii. Casa dei capitelli figurati
x Capitelli figurati, Casa dei, Pompeii
Casa dei capitelli figurati, Pompeii
xx Architecture, Domestic—Italy
Dwellings—Italy
Pompholyx (Disease)
x Dyshidrosis
xx Eczema
Pompilidae *
See Spider wasps
Pon
See Sanka (Social class)
Ponape Island
— Antiquities
sa Nanmatol, Ponape Island
Ponape language *(PL6295)*
xx Melanesian languages
Micronesian languages
Ponca Creek, S.D. and Neb.
xx Rivers—Nebraska
Rivers—South Dakota
Ponca Creek watershed, S.D. and Neb.
xx Watersheds—Nebraska
Watersheds—South Dakota
Ponca Indians *(E99.P7)*
x Ponka Indians

xx Dhegiha Indians
Indians of North America
Siouan Indians
Pond ecology *(Indirect)*
xx Fresh-water ecology
Pond fauna *(Indirect)*
xx Ponds
Zoology
Pond flora *(Indirect)*
sa Bog flora
Marsh flora
xx Bog flora
Fresh-water flora
Marsh flora
Wetland flora
Pond lilies
See Water-lilies
Pond-lily leaf-beetle
xx Water-lilies
Pond smelt *(QL638.O84)*
x Hypomesus olidus
Ponda language
See Lucazi language
Ponden Hall, Eng.
xx Dwellings—England
Pondermotive force
See Ponderomotive force
Ponderomotive force
x Pondermotive force
xx Electromotive force
Ponderosa pine
x Western yellow pine
Pondos *(DT764.P6)*
x Kafirs (African people)
xx Bantus
Ethnology—South Africa
Ponds *(Indirect)*
sa Farm ponds
Pond fauna
Strip mine ponds
xx Water
— China
sa Hua-ch'ing Pond, China
— France
sa Lindre Pond, France
Ponds, Farm
See Farm ponds
Ponds, Fish
See Fish ponds
Pongo pygmaeus
See Orangutan
Pongwe language
See Mpongwe language
Poni language
See Baoulé language
Ponies *(Indirect)* *(SF315)*
sa Basuto pony
Chincoteague pony
Connemara pony
Exmoor pony
Highland pony
Iceland pony
Indian ponies
New Forest pony
Pit pony
Polo ponies
Pony breeding
Pony trekking
Shetland pony
Welsh Mountain pony
Welsh pony
Wild horses
xx Horses
— Stud-books
x Stud-books
xx Horses—Stud-books
Ponka Indians
See Ponca Indians
Ponkapoag, Mass.
See Ponkapog, Mass.

FIGURE 2.5 An excerpt from the *Library of Congress: Subject Headings*

Most of the items in a card catalog are listed three different ways: by author, by title, and by subject. The subject cards provide a handy cross-reference to works related to each other. See Figure 2.6.

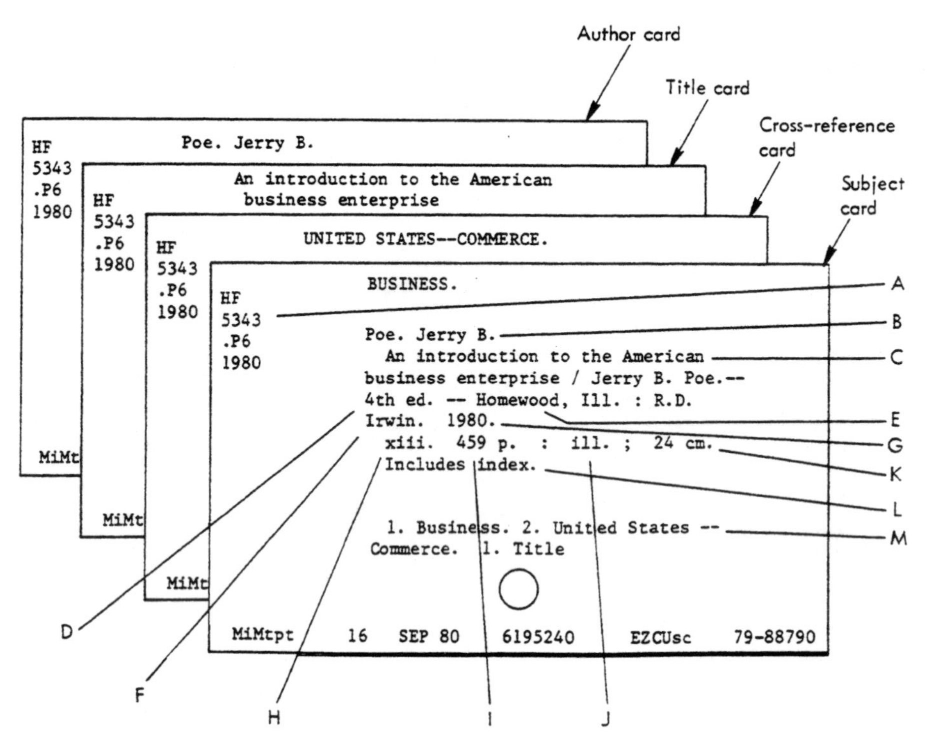

FIGURE 2.6 Card catalog cards

Card Information

The catalog card can give you more information than the title and location of the publications. The subject card in Figure 2.6 shows (A) the call number; (B) the author's name; (C) the title; (D) publishing information; (E) place of publication; (F) the publisher; (G) date of publication; (H) number of preface pages, in roman numerals; (I) number of pages in the book; (J) the presence of illustrations; (K) the size of the book in centimeters (1 centimeter is 0.4 inch); (L) the presence of an index; (M) other cards for this book in the catalog, excluding the author card. The information at the bottom of the card is used by the librarian for ordering the book or a new catalog card.

Suppose you were doing a report on the ape-language question. One book worth reading might be Patterson's book *The Education of Koko*; Koko is the only gorilla in this research. You might look it up in the catalog, write down the call number, and go hunting for it on the shelves. But you may discover that the book is out; it's a very popular book and hard to find. Now you must go back to the catalog to see what else you can find, and perhaps in the meantime you lose the scrap of paper on which you have written the call number for *Koko*. The library, you may decide, is more work than you care for. However, the real problem is in your procedure. A better procedure is to break your search into steps. The first step should be to get as much information as possible from the card catalog. Instead of scribbling down call numbers for a book or two, spend an hour or so constructing a bibliography or list of items; get as many items as you can. List the call number and the *full* bibliographic reference you will need if you use the book in your paper, as in Figure 2.7.

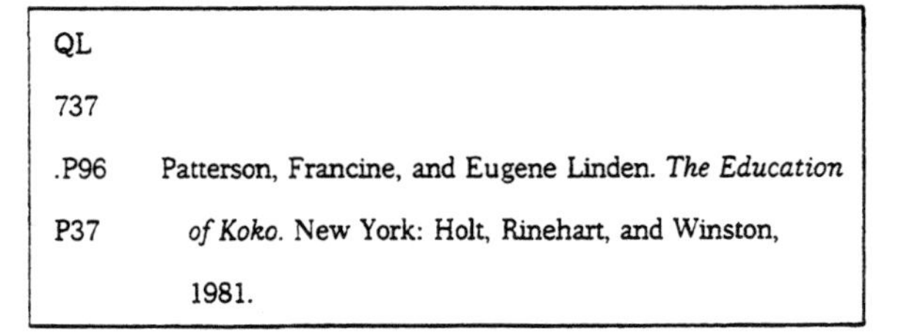
QL
737
.P96 Patterson, Francine, and Eugene Linden. *The Education*
P37 *of Koko*. New York: Holt, Rinehart, and Winston,
1981.

FIGURE 2.7 Write down the call number and the full bibliographic data when you construct the initial bibliography.

Now, when you look for books, you will have all the information you will need later for your bibliography. If you have the call number, you can keep looking for a book until it turns up. In other words, you can reduce the time and frustration of research if you follow a routine for gathering information. You needn't run back and forth to the catalog, redoing steps you have already done. Set yourself the task of constructing a working bibliography (book list) so you have the necessary information to begin with. Then, if—as often happens—the book you want isn't available, you can request that it be held for you when it's returned, or you can request that it be borrowed from another library (interlibrary loan).

Using the *Library of Congress: Subject Headings* to find the correct card catalog descriptors, a researcher might look under "Ape" and "Chimpanzee" before finding the correct descriptor. See Figure 2.8.

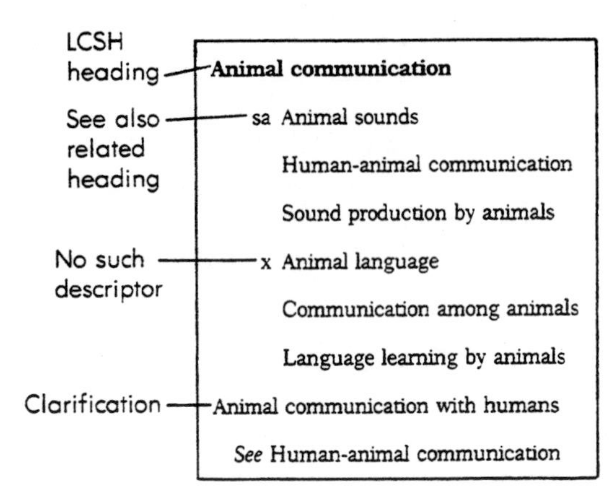

FIGURE 2.8 Excerpt from *Library of Congress: Subject Headings*

Two of these descriptors look promising. If you look up "Language learning by animals" in the card catalog you will find:

> Language Learning by a Chimpanzee: The Lana Project. Duane M. Rumbaugh, ed. New York: Academic Press, 1977.

The other descriptor that seems relevant is "Human-animal communication." Under that heading, the card catalog has the following titles:

> Crail, Ted. Apetalk and Whalespeak: The Quest for Interspecies Communication. Boston: Houghton Mifflin, 1981.
>
> de Luce, Judith and Hugh T. Wilder, eds. Language in Primates: Perspectives and Implications. New York: Springer-Verlag, 1983. [Includes bibliography.]
>
> Linden, Eugene. Silent Partners: The Legacy of the Ape Language Experiments. New York: Times Books, 1986.
>
> Patterson, Francine and Eugene Linden. The Education of Koko. New York: Holt, Rinehart, Winston, 1981. [Includes bibliography.]
>
> Premack, David. ''Gavagai!'' or the Future History of the Animal Language Controversy. Cambridge: MIT P, 1986. [Includes bibliography.]
>
> Sebeok, Thomas A. and Jean Umiker-Sebeok, eds. Speaking of Apes: A Critical Anthology of Two-Way Communication with Man. New York: Plenum, 1980. [Includes bibliography.]
>
> Terrace, Herbert S. Nim. New York: Knopf, 1979.

Savage-Rumbaugh, E. Sue. Ape Language: From Conditioned Response to Symbol. New York: Columbia U P, 1986. [Includes bibliography.]

In just a few minutes with the *Subject Headings* you will save yourself a lot of time and effort at the card catalog, and you may easily find more than enough books to start your research with. Note how many of the items include bibliographies.

ACTIVITY 11 Using one or more bibliographic sources, compile a bibliography of bibliographies. Then use your list to look up these bibliographies. Write down what you find: compile a preliminary bibliography.

STRATEGY THREE: LOOK FOR ARTICLES

In addition to books, the library also holds popular magazines, scholarly journals, and newspapers, all of which contain useful information. To find a book about drug abuse is relatively simple with the card catalog, but how do you find a magazine article or a news report? You can find articles in the "popular press" (general-interest magazines and newspapers) through the appropriate indexes. For example:

Popular Magazines

Alternative Press Index, 1970–.
Covers "alternative" periodicals not found in *Readers' Guide*, radical publications, gay liberation, etc. Most libraries don't carry the periodicals listed in this index.
Essay and General Literature Index, 1900–.
International Index to Periodicals, 1920–65.
Poole's Index to Periodical Literature, 1802–06.
Author Index for Poole's, 1970.
A companion volume, ed. C. Edward Wall. *Poole's Index* lists by title and subject, but not author.
Popular Periodical Index, 1973–.
Covers magazines not in *Readers' Guide*, for example, *Crawdaddy*, *Rolling Stone*, etc.
The Readers' Guide to Periodical Literature, 1900–.
Standard Periodical Directory.
Lists titles of periodicals. A good source if you want to find out whether there are periodicals in your area of research.
Ulrich's International Periodicals Directory, 1973–.

General Interest Newspapers

Bell and Howell's Index to the Christian Science Monitor, yearly.
Facts on File: A Weekly News Digest, 1940–.
Contains current events covered in newspapers.
London Times Official Index, 1906–.
Newspaper Index.
[Chicago Tribune, Los Angeles Times, New Orleans Times-Picayune, Washington Post]
New York Times Index, 1913–.
Palmer's Index to The [London] Times, 1790–1905.
Wall Street Journal Index, 1958–.
Washington Post Index.

The most generalized index is the *Readers' Guide to Periodical Literature*, which covers many popular magazines, starting with the year 1900. In the index you will find articles listed by title, by author, and by subject matter, and also extensively cross-indexed under many different descriptors. Even if you have only the fuzziest notion of where to start, the indexes will quickly give you a wealth of information. Check the introductory pages of any volume of the index to find out what abbreviations and symbols are used in the entries.

You can see that the *Readers' Guide* can offer many ideas for research topics. It can also offer alternate topics when you don't find sufficient material. The short excerpt in Figure 2.9 contains a great deal of information:

1. Under "Medical news" you will find a continuing ("cont") feature: "Latest in health," with illustrations (il) in *U.S. News*, volume 88. The volume number covers all the issues and is therefore not repeated. The feature appeared each month except August. The numbers "88:70 F 4" mean volume 88, page 70, February 4. The year of the most recent article is given at the end, "'81."
2. Under "New advances in medicine" note a similar continuing feature by the author L. Galton in the magazine *50 Plus*.
3. Under "Medical records" you will find the title of an article, "Protecting Medical Records," by the author G. Odening in *Forbes* magazine, volume 126, page 165 plus (indicating additional pages), December 8, 1980.
4. Under "Mediterranean Sea" there is a cross-reference to "Petroleum—Mediterranean Region."
5. "Medoff, James" is the subject of an interview (see the bracketed information). The abbreviation "pors" means "portraits" (pictures of people, not the same as "il").
6. "Medwick, Cathleen" is the author of an interview with Norman Mailer. The article is cross-referenced under "Mailer, Norman."
7. "Meehan, Francis X." is the author of "Disarmament in the Real World," *America* 143, pages 423–426, December 27, 1980.

MEDICAL news
Latest in health [cont] il U.S. News 88:70 F 4; 63 Mr 10; 84 Ap 14; 88 My 19; 70 Je 23; 74 Jl 21; 89:42 S 1; 91 O 6; 113 N 17; 87 D 29 '80–Ja 5 '81
New advances in medicine [cont] L. Galton. 50 Plus 20:34–5 Mr; 48–9 Ap; 40 Je; 22–3 Ag; 32–3 O; 10 N '80
MEDICAL records
Protecting medical records. G. Odening. Forbes 126:165+ D 8 '80
MEDITERRANEAN Sea
See also
Petroleum-Mediterranean Region
MEDOFF, James
Would you believe unions are good for productivity? [interview by S. Kinsley] pors Fortune 102:149–50+ D 1 '80
MEDWICK, Cathleen
(int) See Mailer, Norman. Norman Mailer on love, sex, God, and the devil
MEEHAN, Francis X.
Disarmament in the real world. America 143: 423–6 D 27 '80
MEESE, Edwin, 3d
Reagan does it his way. D.M. Alpern and others. il por Newsweek 96:10–11 D 29 '80
MEEUS, Jean. See Goffin, E. jt auth.
MEGALITHIC monuments
Megalithic monuments [Neolithic period] G. Daniel. bibl(p 162) il Sci Am 243:78–81+ Jl '80; Discussion. 243:6 N '80
MEINKE, Peter
Twisted river [story] il Atlantic 246:80–4 D '80
MELLOW, James R.
Transcendental admirers: Turner's American Friends. il Art News 79:80–3 D '80
MEMBRANES (biology)
Insulin receptors: differences in structural organization on adipocyte and liver plasma membranes. L. Jarrett and others. bibl f il Science 219:1127–8 D 5 '80
MEMORY
See also
Eidetic memory
Remembrance of times lost [study of theory of permanent memory and recall by Elizabeth and Geoffrey Loftus] Psychol Today 14:98–100 N '80

FIGURE 2.9 An excerpt from the *Readers' Guide*, January 1981

8. "Meese, Edwin, 3d" (Edwin Meese, III) is the first author of an article by Meese, D. M. Alpern, and others.
9. "Meeus, Jean" is the co-author ("jt auth") of an article that is indexed under the other author's name, Goffin, E.
10. "Megalithic monuments" is the subject and the title of an article by G. Daniel in *Scientific American*. Note that this article has a bibliography ("bibl") on page 162. Daniel's article in July is followed by another article or a response ("Discussion") in November.
11. "Meinke, Peter" is the author of a fictional account ("story") called "Twisted River."
12. "Membranes (biology)" is the general subject heading for "Insulin Receptors: Differences in Structural Organization on Adipocyte and Liver Plasma Membranes," by L. Jarrett and others in *Science*, volume 219, pages 1127–28, December 5, 1980. Note this article has a bibliography.

Note that while *Science* and *Scientific American* are technical journals and not really aimed at the general public, they are nonetheless indexed in the *Readers' Guide*.

Just as with the card catalog, there are routines to follow with the indexes. Unless you make yourself sit down and work on your bibliography first, you will find yourself running back and forth between the indexes and the bound volumes of periodicals.

The indexes are bound chronologically, and many of them cover several decades. You will discover that a broad subject like "drugs" will produce hundreds of articles—far too many for you to read., Thus you will save yourself even more time if you limit your search to some small aspect of the overall subject. If you were interested in just a psychological profile of marijuana smokers, you could ignore many of the articles in the indexes. Even so, you will still find many pertinent articles. Now how should you proceed?

Start Where You Are

Unless you are doing a historical study (marijuana in the 1920s, for example), there is a general rule that all research starts where you are. That is, you start with the most recent research and work your way back to older research, going as far back as your study requires. You might limit yourself to research done in the last five years or the last ten years, for example. Research is never haphazard; it doesn't make sense to jump in at an arbitrary point. Furthermore, recent research is based on previous research. This means you can find the bibliographical material you want by finding the most recent research. Finally, there is a possibility that any conclusions you reach based on old data may be invalidated or subsumed by more recent research you could have read.

ACTIVITY 12 Use some newspaper and magazine indexes to find several articles related to your research topic. Collect a preliminary bibliography of periodical articles.

Professional, Technical, and Specialty Journals

In addition to the general magazine and newspaper indexes, there are many specialized indexes covering professional journals and other information not generally found in popular magazines. For example:

ACCOUNTING
Accountant's Index, 1921–.
AGRICULTURE
Agricultural Index, 1919–64.
Yearbook of Agriculture, 1894–.
Covers new subject each year.
ANTHROPOLOGY
Anthropological Index.

ART

Art Index: A Cumulative Author and Subject Index to a Selected List of Fine Art Periodicals, 1929–.

Index to Art Periodicals, 1962–.

BIOLOGY

Biological Abstracts, 1926–.

Biological and Agricultural Index, 1964–.

Previously titled *Agricultural Index*, 1919–64.

BioResearch Index.

BUSINESS

Business Periodicals Index, 1958–.

Accounting, advertising, banking and finance, labor, etc. Previously *Industrial Arts Index*, 1913–58.

CHEMISTRY

Chemical Abstracts, 1907–.

CLASSICS

International Guide to Classical Studies.

COMPUTERS

Computer Abstracts, 1957–.

Computer Literature Index, 1971–.

CRIMINOLOGY

Criminology Index.

DISSERTATIONS

American Doctoral Dissertations.

Lists all North American dissertations. If you know the author or the title, you can look it up. *A.D.D.* lists but does not describe dissertations.

Comprehensive Dissertation Index, 1861–1972.

There are 37 volumes. The first 32 volumes are devoted to subject areas: 1–4 chemistry, 5 math and statistics, 6 astronomy, and so forth. Use a key word search in the appropriate volume. Volumes 33–37 are the *Author Index*.

After 1973 there are yearly supplements to the index: two volumes for the natural sciences, two volumes for social sciences and humanities, one volume for authors. The *Comprehensive Dissertation Index* is always current up to the preceding year, in monthly supplements.

Note that *A* and *B* in dissertation references identify humanities and social sciences (A) and natural sciences (B).

DATRIX II, 1967–.

Direct Access to Reference Information: A Xerox Service. This is a computer database to dissertations and bibliographical information. You can ask DATRIX *whether* there have been any dissertations about frogs, for example.

Dissertation Abstracts International, 1970–.

originally *Microfilm Abstracts*, 1938–51.
renamed *Dissertations Abstracts*, 1952–69.
600-word abstract on each dissertation.

DRAMA
American Drama Criticism, 1890–.
Cumulated Dramatic Index, 1909–40, 1965.
Play Index, 1953–.

ECONOMICS
Index to Economics Journals, 1886–1963.
7 volumes, 1966–.

EDUCATION
British Education Index.
Current Index to Journals in Education, 1969–.
Education Index, 1929–.
Resources in Education, 1956–.
State Education Journal Index, 1963–.

ENERGY
ERDA Energy Research Abstracts.

ENGINEERING
Engineering Index, 1884–.

ENVIRONMENT
Environmental Index, 1971–.
Pollution Abstracts, 1970–.

ETHNIC
Index to Literature on the American Indian, 1970–.
Index to Periodical Articles by and about Negroes, 1950–.

FILM
Film Literature Index, 1974–.
International Index of Film Periodicals, 1973–.
New York Times Film Reviews, 1913–68.
Updated with supplements.
Retrospective Index to Film Periodicals, 1930–71.

HUMANITIES
Humanities Citation Index.
Humanities Index, 1974–.
Previously titled *Social Sciences and Humanities Index*, 1965–74, and *International Index*, 1907–65.

INDUSTRIAL ARTS
Industrial Arts Index, 1913–57.
divided, renamed *Applied Science and Technology Index*, and *Business Periodicals Index*, 1958–.

LAW
Index to Legal Periodicals, 1908–.

LINGUISTICS
Analecta Linguistica.
Linguistics and Language Behavior Abstracts.

LITERATURE

Abstracts of English Studies, 1958–.

Index to Little Magazines.

MLA International Bibliography, 1921–.

Nineteenth Century Readers' Guide to Periodical. Literature, 1800–99.

Largely literary.

T.L.S.: Essays and Reviews from the Times Literary Supplement, 1963–.

MEDICAL

Index Medicus, 1959–.

Hospital Abstracts.

MUSIC

Music Index: The Key to Current Music Periodical Literature, 1949–.

Popular Music Periodical Index.

RILM Abstracts, 1967–.

PHILOSOPHY

Philosopher's Index: An International Index to Philosophical Periodicals, 1967–.

PHYSICS

Current Physics Index, 1975–.

Physics Abstracts, 1898–.

PSYCHOLOGY

Annual Review of Psychology, 1950–.

New developments in psychology.

Psychological Abstracts, 1927–.

Previously covered by *Psychological Index*, 1895–1936.

RELIGION

Catholic Periodical Index, 1939–.

Christian Periodical Index, 1956–.

Index to Religious Periodical Literature, 1949–.

New Testament Abstracts, 1956–.

Religious and Theological Abstracts, 1958–.

SCIENCE AND TECHNOLOGY

Applied Science and Technology Index, 1958–.

Previously titled *Industrial Arts Index*, 1913–57.

General Science Index, 1978–.

Science Abstracts, 1889–.

Science Citation Index.

SOCIAL SCIENCES

Social Sciences Citation Index.

Social Sciences Index, 1974–.

Previously titled *Social Sciences and Humanities Index*, 1965–74, and, *International Index*, 1907–65.

Sociological Abstracts, 1952–.

SPEECH
Speech Index, 1900–65.
WOMEN'S STUDIES
Women's Studies Abstracts.

ACTIVITY 13 Use the specialized indexes to find articles related to your research thesis in professional journals. Compile a preliminary bibliography of professional journal articles.

Other Sources

ALMANACS
Economic Almanac, 1940–.
Business and industry facts and figures.
The World Almanac and Book of Facts, yearly.
COLLEGES
American Universities and Colleges, 1964–.
Comparative Guide to American Colleges, 1965–.
Education Directory, 1912–.
GPO.
Lovejoy's Complete Guide to American Colleges and Universities, 1940–.
Information about schools, faculty/student ratios, etc.
Patterson's American Education, 1904–.
GROUPS
Encyclopedia of Associations.
Groups, clubs, federations, organizations.
Encyclopedia gives names and addresses.
MANUFACTURING
Thomas Register of American Manufacturers, 1905–.
Products and the companies that make them.
NEWSPAPERS
Editor and Publisher International Yearbook.
A listing and detailed description of daily and weekly newspapers in the United States and Canada. Includes such things as names of editors, reporters, owners, and so on. A good source if you are looking for "morgue" material that may be available in local newspapers.
Ayer's Directory of Newspapers and Periodicals, 1880–.

Find out as much as you can about your library's indexes. You will discover that they can give you dozens of articles on practically any subject—one reason for limiting yourself to the smallest possible aspect of any subject. The broader the subject, the more books and articles you will have to read and assimilate into your subject. Note the excerpt from the *New York Times Index* in Figure 2.10.

BOLLS, Richard. See also Boarding Houses, Ja 24
BOLOGNA, Joseph. See also theater—Revs, Ja 28
BOLOGNA (Italy). See also Travel—Italy, Mr 22
BOMB Shelters. Use Air Raid Shelters
BOMBECK, Erma. See also Television—Programs, Mr 12
BOMBS and Bomb Plots. See also Airlines—Accidents etc,
Air Florida, Mr 18. Airplanes—US—Mil Aircraft, Ja 13, 14, 18, Mr 25. Armament. Arms Control. Atomic Energy. Banks—US, Mr 22. Brazil, Mr 27. Chemical Warfare, Ja 18. Corsica, Ja 31, F 13, 14, 18, Mr 26. Costa Rica, Mr 18, 22, 31. Courts—NYS—Supreme Court (State), Ja 24, Mr 21. Cuba, Mr 17. Elec Light—Salvador, El, F 7. Football—Bowl Games, Super Bowl, Ja 22. France—Pol, Ja 4. Gambling—NJ, Mr 9, 10. Gambling—NYS, Mr 31. GB—Pol, Ja 8, 9, 10. Guadeloupe, Ja 5, 11, 31, F 19. Honduras, Mr 27. Hong Kong, F 22. Hotels etc—Kenya, Ja 1, 2, 3, 4, 5, 6, 8, 9. Hotels etc—NJ, Mr 10. Iran, F 5. Ireland, Northern, Ja 9, 10, 27, F 2, 8. Italy—Pol, Ja 7, Mr 22. Jews—France—Anti-Semitism, F 17. Jews—NYC Met Area, Mr 6. Jews—Uruguay—Anti-Semitism, F 26. Middle East— Israeli-Arab Conflict, Ja 10. Missiles. Murders—Pa, Mr 16. News—Brazil, Mr 27. News—Cyprus, Ja 4. NATO, F 13, 15, 17, 24. Oil—Salvador, El, F 3. Pakistan, F 17, Mr 6. Philippines, F 15, Mr 20. Puerto Rico, Ja 13, 18, F 5. Retail Stores—Australia, Ja 14. Robberies—NYC, Quartermaster Pinball Parlor (NYC), Mr 25. RC Ch—Pope, F 17, Mr 6. Salvador, El, F 3, 7, 26, Mr 10, 22. Ships—Accidents etc, Nellie M (Freighter), F 8. Somalia, Ja 31. Spain, Ja 20, 21, 26. Syria, Ja 4. Television—Radio Free Europe, F 22, 23. Transit Systems—Pa—Philadelphia (Pa), Mr 23. Turkey, Ja 5. USSR—Pol, F 20. US—Pol—Fringe Pol Movements, Mr 18, 19. US Armament—Defense Contracts, Mr 21. Yemen, People's Dem Repub, F 17. Yugoslavia, Ja 24, Mr 17, 28, War names
2 bomb devices are found in basement of Harlem, NYC, apartment house; 2d was found by bomb squad detectives as they were dismantling 1st; illustration (M), Mr 17, II, 2:5
BONADONNA, Gianni (Dr). See also Cancer, Ja 1
BONANNO, Joseph Sr. See also Crime—US, Ja 13
BOND, Elaine R. See also Chase Manhattan Bank (NYC), F 3
BOND, James A. See also Air Conditioning, F 28
BOND, Langhorne M. See also Airlines—Accidents etc, F 2
BOND, Mimi. See also Cooking, Ja 4
BOND Corp. See also subsidiaries, eg, Endeavour Resources Ltd
BOND, Beulah
Dies at age 92; portrait (S), Ja 13, IV, 19:3
BOOK Fairs. See Books
BOOK Reviews. Note: There are 2 listings below: 1st a listing by author for works by a single author; 2d a title listing for all other works, eg, anthologies, collaborations etc

- A -

Abish, Walter: How German Is It reviewed by Betty Falkenberg, Ja 4, VII, p8
Adams, Douglas: The Hitchhiker's Guide to the Galaxy reviewed by Gerald Jones, Ja 25, VII, p24
Aldiss, Brian W: An Island Called Moreau reviewed by Christopher Lehmann-Haupt, F 17, III, 20:1
Alexander, Karl: A Private Investigation reviewed by Newgate Callendar, F 1, VII, p31
Aliki: Digging Up Dinosaurs reviewed by Susan Bolotin, Mr 8, VII, p30
Ames, Mildred: The Dancing Madness reviewed by Joyce Milton, F 1, VII, p28
Amesbury, James Edward: Sporting Chance reviewed by Newgate Callendar, Ja 4, VII, p17
Anderson, Patrick: High in America: The True Story Behind Normal and the Politics of Marijuana reviewed by Christopher Lehmann-Haupt, F 26, III, 18:1
Anson, Robert Sam: Gone Crazy and Back Again: The Rise and Fall of the Rolling Stone Generation reviewed by John Leonard, F 10, III, 11:1
Arbeiter, Jean S: Womanlist reviewed by Susan Jacoby, Mr 22, VII, p15
Arensberg, Ann: Sister Wolf reviewed by Christopher Lehmann-Haupt, Ja 15, III, 21:4
Aries, Philippe: The Hour of Our Death reviewed by John Leonard, F 16, III, 17:5; by Robert Nisbet, F 22, VII, p7
Asimov, Isaac (Dr): In the Beginning reviewed by Timothy Ferris, Mr 15, VII, p16
Auchincloss, Louis S: The Cat and the King reviewed by Doris Grumbach, Mr 15, VII, p14; by John Leonard, Mr 16, III, 15:1
Avi: A Place Called Ugly reviewed by Bryna J. Fireside, Mr 1, VII, p24

- B -

Bailey, Anthony: America, Lost & Found reviewed by Christopher Lehmann-Haupt, Ja 29, III, p17; by John Russell, F 8, VII, p12

FIGURE 2.10 An excerpt from the *New York Times Index*

The headings in the *New York Times Index* identify subject matter, not headlines or the titles of articles. Occasionally an entry will contain an abstract of the article, followed by date, section, page, and column. The *Index* uses the symbols *S*, *M*, and *L* to indicate article length: *S* means "short," half a column or less; *M* means "medium," up to two columns; *L* means "long," over two columns.

To find recent magazine articles use The Magazine Index, which, in addition to being more comprehensive than the *Readers' Guide*, is electronic and a good deal easier to use. (See "The Magazine Index," pp. 49–51.)

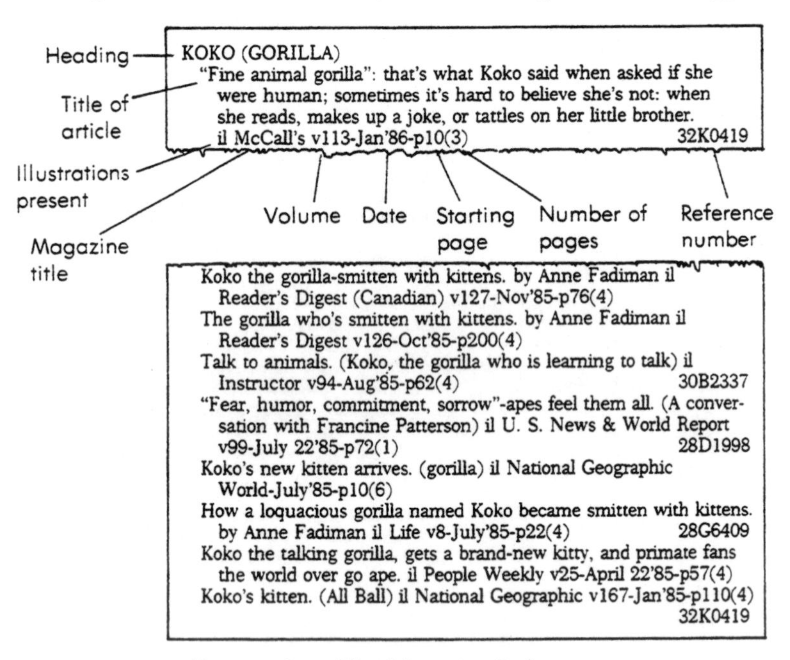

KOKO (GORILLA)
"Fine animal gorilla": that's what Koko said when asked if she were human; sometimes it's hard to believe she's not: when she reads, makes up a joke, or tattles on her little brother.
il McCall's v113-Jan'86-p10(3) 32K0419

Koko the gorilla-smitten with kittens. by Anne Fadiman il Reader's Digest (Canadian) v127-Nov'85-p76(4)
The gorilla who's smitten with kittens. by Anne Fadiman il Reader's Digest v126-Oct'85-p200(4)
Talk to animals. (Koko, the gorilla who is learning to talk) il Instructor v94-Aug'85-p62(4) 30B2337
"Fear, humor, commitment, sorrow"-apes feel them all. (A conversation with Francine Patterson) il U. S. News & World Report v99-July 22'85-p72(1) 28D1998
Koko's new kitten arrives. (gorilla) il National Geographic World-July'85-p10(6)
How a loquacious gorilla named Koko became smitten with kittens. by Anne Fadiman il Life v8-July'85-p22(4) 28G6409
Koko the talking gorilla, gets a brand-new kitty, and primate fans the world over go ape. il People Weekly v25-April 22'85-p57(4)
Koko's kitten. (All Ball) il National Geographic v167-Jan'85-p110(4) 32K0419

FIGURE 2.11 Excerpt from The Magazine Index

Animal communication
See also
Animal sounds
Human-animal communication
Pheromones
Alarm call responsibility of Mallard ducklings: the inadequacy of learning and genetic explanations of instinctive behavior D. B. Miller and C. F. Blaich. bibl *Learn Motiv* 15:417–27 N '84
Continuity between primate comunication and human speech? review article. J. W. Froelich. *J Anthropol Res* 40:597–602 Wint'84

FIGURE 2.12 Excerpt from *Social Sciences Index*, April 1985 to March 1986

Under the main heading "Animal communication" you will find a review article by Froelich. Review articles are especially good because they analyze research and discuss strengths and errors in experiments, and assumptions in the research. Note the *See also* leading to yet another article.

Human-animal communication

A vocabulary test for chimpanzees (Pan troglodytes). R. A. Gardner and B. T. Gardner. bibl il *J Comp Psychol* 98;381–404 D '84

FIGURE 2.13 Cross-reference from *Social Sciences Index*

Abbreviations and symbols used in the indexes are explained in the front pages of each index: *J Anthropol Res* (*Journal of Anthropological Research*); *J Comp Psychol* (*Journal of Comparative Psychology*).

Since this research topic deals with language, it will be good to look for an index that also deals with language, like *Linguistics and Language Behavior Abstracts*. An index like this not only gives bibliographic references, it gives an abstract for each reference.

5800	NONVERBAL COMMUNICATION	Abstract	Page
5810	Human Nonverbal	7771	1301
5811	Animal Communication	7791	1303
5812	Art as Language	7796	1304

8507794

Muncer, Steven J. (Nassau Community College, Garden City NY 503), Is Nim, the Chimpanzee, Problem Solving?, Perceptual and Motor Skills, 1983, 57, I, Aug, 132–134.

Data concerning lang acquisition by the chimpanzee Nim, who was taught to use American Sign Language (Ameslan), are reviewed. It is argued that the importance of the project has been overemphasized. Many flaws in the project are noted. Nim learned only what he was trained to do, and had little opportunity to really learn or use lang. 9 References. Modified HA.

8507795

Premack, David (U Pennsylvania, Philadelphia 19104), **Possible General Effects of Language Training on the Chimpanzee**, UM *Human Development*, 1984, 27, 5–6, Sept–Dec, 268–281.

Once the chimpanzee has been exposed to lang training, it solves certain kinds of problems that it does not solve otherwise. Specifically, it can solve problems on a conceptual rather than a sensory basis. It is only after it has been lang trained that the normal chimpanzee can match, ie, 3/4 apple with 3/4 cylinder of water, ie, match equivalent proportions of objects that do not look alike. Similarly it can match not only relations of sameness [. . . .]

FIGURE 2.14 Excerpts from *Linguistics and Language Behavior Abstracts*, Dec. 1985, Vol 19, No 4

STRATEGY FOUR: USE LIBRARY COMPUTERS, ERIC, MICROFILM INDEXES, AND MICROFORM READERS

Computer Terminals

Many large libraries today use computer terminals, a fact that is becoming increasingly helpful to many researchers. A number of systems are used across the nation, but they are all quite similar from the user's point of view. For purposes of illustration, this discussion refers to the oldest of the systems, OCLC (Online Computer Library Center).

A large central computer serves many distant libraries and many computer terminals. If you have an OCLC terminal in your library, you can find books and periodicals not only in your own library but elsewhere as well. As a researcher, you can find material if you have either the title, the author's name, the LC (Library of Congress) number, or the international standard book number (ISBN). You can't use the OCLC system directly for a subject search; that is, you can't use the system to find material on general subjects (abortion, defense spending, and so on). But it does provide quick access to the complete bibliographic information for more than 15 million holdings of over 3000 libraries. It appears certain that the computer will one day replace the card catalog.

Library computer systems were designed primarily for the librarians' use; thus the systems are capable of more services than students require. At first sight the terminal may look complex, but it's quite simple to use. The terminal is an expanded typewriter keyboard and video screen. To find a book, you need only type in an access number (each subscribing library has its own access number), then type in the title of the book. If you type in the author's name instead of the title of the book, the computer will display all the books written by that author. In some cases, where there are books with the same or similar titles (for example, *Basic Chemistry*) by different authors, give both the author and the title.

Instead of the full title or the full name of the author, the computer accepts only a few letters of each word. For example, the pattern for a title might be 3,2,2,1—which means that you are to type in only the first three letters of the first word in the title, the first two letters of the second word, the first two letters of the third word, and the first letter of the fourth word. Type commas to separate groups of letters. Thus, to type in the title for Rapoport's *Fights, Games, and Debates*, you would type the entry:

F I G, G A, A N, D

Then you must press two keys: DISPLAY/RECORD and SEND. The record for the book will then appear on the screen, showing the card catalog information for the book, including the call number, along with a note telling you whether your library has a copy of the book.

Many libraries also offer computer searches of various subject-oriented databases. Many of these databases are computerized versions of the printed indexes listed earlier, such as *Psychological Abstracts* or *Current Index to Journals in Education.* Working with a reference librarian, you may be able to order, usually for a small fee, a computer-generated bibliography tailored to your specific research question. Ask one of the reference librarians in your library to describe the particular services available to you.

ERIC

In addition to a general computer system like OCLC, your library may have a specialized system like ERIC. *E*ducational *R*esources *I*nformation *C*enter (ERIC) is a clearinghouse that provides access to educational materials, particularly unpublished documents. Data concerning teaching and various subjects related to students may be available through ERIC. ERIC documents are listed in two cumulative indexes: *Resources in Education* and *Current Index to Journals in Education.* Before using either index, find your subject descriptors in the *Thesaurus of ERIC Descriptors* for many different possible descriptions (movies, for example, might be listed under *film, cinema, motion pictures, moving pictures, art, Hollywood, entertainment,* or *theater*). Look up these descriptors until you find items that sound promising; write down the ED (Educational Document) numbers and the titles of the documents. You can then do one of three things: Read the document in its original journal (if your library has it); read an abstract of the document in the Resumé section of *Resources in Education*; or order a microfiche or a hard copy of the document through ERIC. Many libraries offer computer searches of the ERIC database.

Microfilm Indexes

The microfilm indexes represent an advance over the printed indexes most students know. Microfilm indexes tend to be much larger, much more comprehensive than the print indexes; they offer superior coverage, at least for the most recent articles. But their chief advantage is that they permit researchers to search electronically. Moving at very high speeds, faster than anyone can read, microfilm indexes can scroll through alphabetically arranged citations much faster than an unaided researcher can thumb through printed pages of citations.

The Magazine Index

The Magazine Index covers 354 of "the most popular magazines in America." The coverage is very complete; the index covers everything in the magazines that could conceivably be of research interest, meaning virtually everything but the advertisements. It subsumes the *Readers' Guide* (that is, *Readers' Guide* coverage is included) and also includes many other magazines. Coverage began in 1976; each microfilm reader contains an index covering approximately the last five years; it is current except for the previous month or two. (See Figure 2.11.)

In addition to news and feature articles, you can find reviews of books, plays, films, and other performances listed under the title of the performance. The Magazine Index shows the reviewer's overall judgment as a letter grade: *A*, excellent; *B*, good, a welcome addition; *C*, OK, competent; *D*, don't bother, not very good; *F*, terrible.

EXAMPLE REVIEW

On Golden Pond. (Apollo Theatre, NY) Theater Reviews. rev. by Gill, Brendan.
A New Yorker V55 March 12 '79 p. 107 (2)

Note that The Magazine Index has its own style and dispenses with such matters of standard format as quotation marks, italics, and some punctuation. This reviewer gave the performance an excellent rating (note the grade, A, after the reviewer's name). This is a review of a theater performance, not to be confused with the film of the same title. Page numbers at the end of the item indicate the beginning page and the number of pages (in parentheses) of the review.

Suppose you were looking for film reviews to use in your paper on *Superman II*. If you looked up the title in The Magazine Index, you would find it listed as a heading:

Superman II (Moving-Picture-Reviews)

Under this heading, you would find dozens of reviews listed between April and December 1981. For example:

B+ Ladies Home Journal V98 October '81 p. S6 (1)
B– Penthouse V12 August '81 p. 46 (1) rev by Kael, Pauline.
A New Yorker V57 July 13 '81 p. 81 (2)

Many short reviews are listed, but a review by a noted critic is identified for you (the review by Pauline Kael in the example).

In addition to reviews, you would find articles about the film: descriptive news articles, publicity releases, reports about the picture's premiere, public reception, and so on. For example:

What makes a film fly in Japan.
Business Week June 29 '81 p. 97 (1)

Furthermore, still other articles—giving criticism, interpretation, and so forth—are listed:

> Christopher Reeve—Superman Who's Only Human. by Tavris, Carol il portrait Mademoiselle V86 Oct '80 p. 72 (3)

And there are other articles still, covering the economic aspects of the film, marketing, television uses, and so on. Thus you will find a very thorough indexing of nearly anything having to do with the film. College students looking for recent magazine articles will find The Magazine Index the quickest and easiest index to use. (See Figure 2.11.)

"Hot Topics"

A spinoff of The Magazine Index is "Hot Topics," a monthly printout issued in a loose-leaf binder, covering available articles on currently popular (in the news) issues such as AIDS, Star Wars, the President's health, unemployment, and so on. This print material is frequently kept near the microfilm readers. In effect, many current issues have been preresearched in "Hot Topics."

The National Newspaper Index

The National Newspaper Index covers five major papers: the *New York Times, Christian Science Monitor, Los Angeles Times, Washington Post,* and *The Wall Street Journal.* It starts with January 1, 1979; each microfilm reader contains indexing for the latest three years. The entries show title, author, number of column inches, name of paper, date, year, section (if applicable—the *Monitor* doesn't have sections), page numbers, and column. For example:

> Superman II (Moving-Picture-Reviews)
> rev by Canby, Vincent 10 col. in B+ NYT June 21 '81 sec 2 p. D39 col. I

Note that reviews give grades here, too, as in Canby's B+ grade for this film.

The Business Index

The Business Index covers 357 magazines, journals, and similar publications; the *Wall Street Journal; Barrons;* and the financial section and other relevant business articles in the *New York Times.* For example:

> LIMITED GROWTH POLICY
> Our property rights are being eroded; its time for a careful examination of no-growth and limited-growth legislation.
> by Pat Craig Real Estate Today V12 Oct '79 p. 26 (7) see also
> Land use

The Legal Resource Index

The Legal Resource Index covers 680 law journals, six law newspapers, and relevant law materials from the *Library of Congress Monograph and Government Publication Catalog* (LCMARC). It also contains relevant material from The Magazine Index and The National Newspaper Index.

Materials are indexed by subject, by author and/or title, by cases, and by statutes. Those with asterisks are the journals listed in The Legal Resource Index.

SUBJECT LISTING

Government Spending Policy

*The Regional pattern of federal fund flows. by Allen D. Manvel. 12 Tax Notes 30–32 Jan 5 '81

AUTHOR LISTING

Manvel, Allen D.

*The Regional pattern of federal fund flows. by Allen D. Manvel. 12 Tax Notes 30–32 Jan 5 '81

LCMARC entries in The Legal Resource Index are in the following formats:

SUBJECT LISTING

Accident Law

By accident, not design: -the case for comprehensive jury reparations/-Eli P. Bernzweig. New York, N.Y.: Praeger. c 1980 xi, 221 p.; 25 cm 80020815

(The number at the end, 80020815, is the Library of Congress number.)

CASE LISTING

United States, Laird V

391 F. Supp. 656 (N.D. Ga. 1977) Recent tax developments, regarding purchase of sports franchises—the game isn't over yet. by James F. Ambrose 59 Taxes 739–762 Nov '81

STATUTE LISTING

Internal Revenue Code

I.R.C. 501 (c) (3) Public Interest law firms and client-paid fees. by Terrence J. McCartin 33 Tax Law. 915–956 Spr '80.

Library Databases

As we move into the high-tech era of computers and laser printers, libraries too are increasingly acquiring electronic aids to research. Microfilm indexes such as The Magazine Index and The Newspaper Index will soon be replaced with true databases that you will be able to search by subject headings.

The InfoTrac Database will permit researchers to search for periodical articles in a database at their own library. Database searches are already available to researchers with modems that can access Dialog, Knowledge Index, and similar databases for magazines, professional journals, newspapers.

Databases tend to be huge, so they may not be the best source if you are looking for something simple, like recent articles on Cambodia, AIDS, and so forth. General categories will be easy to find in printed indexes. But for more specific searches, like drug abuse *and* adolescents (a specific subcategory, not the whole drug issue), computerized databases are by far the superior source.

But using modems through the telephone lines can be an expensive proposition. You have to pay for the telephone call—frequently long distance—as well as paying an access fee to the database service. In contrast, the new technology, already installed in some libraries, allows libraries to purchase entire databases on videodiscs. Thus students can access their own school's mainframe through library terminals (multiple access—the number of terminals doesn't affect the speed of access). Furthermore, each terminal is equipped with a printer so students can not only search the database but also get a printout of the result.

To date, the videodisc holds a half-million citations to articles in about 1000 business, general interest, and legal periodicals and several newspapers. Citations begin with January 1982 and are updated monthly.

The whole system is simple to use, and students in the test sites have responded very favorably.

Microform Readers

You may already be aware of the microform readers in your library. *Microform* is a general term covering both microfilm and microfiche. Microfilm is wound on a reel or spool and looks somewhat like motion-picture film. Microfiche is a small sheet of film material, similar to a postcard in size and shape. An entire issue of a newspaper or magazine can often be reproduced on a single spool of microfilm or a single microfiche.

The microforms are used to store back issues of newspapers and some editions of magazines. ERIC documents and many other kinds of books, pamphlets, and other materials are also available on microform. To find an article on microform, first use the appropriate index—for example, the *New York Times Index*. The index will give you the date on which articles related to your topic appeared. Using the dates from the index, you can get the proper microfilm spools.

Microfilm is threaded into a reel-to-reel microfilm reader, which magnifies the print to normal size. Microfiche is loaded on a table-loading reader. On either kind of machine, you use the controls to move the microform back and forth to find the page you are looking for. When you are working with a newspaper, you must be careful to advance to the right section of the paper, then you can move the page up and down to find the specific article.

Some microform readers are equipped with photocopiers so that you can get a photocopy of an article in microform.

ACTIVITY 14 Find the microfilm issue of the *New York Times* for the date of your birth; write a brief synopsis of newsworthy events from that day.

STRATEGY FIVE: LOOK FOR REPORTS, OTHER SPECIALIZED INFORMATION

Government Documents, Reports

Congressional Directory, 1809–.
A guide to the government: Congress and its committees; the courts and judges; agencies and officers of the executive branch: names, addresses, phone numbers, and the like. Researchers seeking information from any branch of government must first find the appropriate person or agency.

Congressional Record Index.
Compiled after each session of Congress; there is no cumulative index. To use it, you must know when a subject was discussed in Congress. The *Congressional Record* contains verbatim (sometimes edited or "expanded") transcripts of what was said on the floor of Congress and the Senate each day.

Index to Current Urban Documents.

Population Index.

Public Affairs Information Service Bulletin, 1915–.
Government documents and related publications.

Statesman's Year-Book, 1864–.

United States Government Organization Manual.
A more detailed companion to the *Congressional Directory*.

United States Government Publications: Monthly Catalog, 1895–.
Selected List of U.S. Government Publications.
A shorter version of the *Monthly Catalog*.

Subject Bibliography Index.
From GPO a list of bibliographies of available material.

Vital Speeches of the Day.

Statistical Information

Statistical Abstract of the United States, 1879–.
Prepared by the U.S. Bureau of the Census: GPO. Facts and figures about America and Americans.

Biographical Sources

Check on experts, identify researchers. There are a number of sources you can check, but it's difficult to find biographies on living people, especially if they aren't celebrities.

American Men and Women of Science, 1972–.
Previously *American Men of Science*.
Biography Index, 1946–.
Contemporary Authors, 1962–.
Living authors, fiction and nonfiction.
Current Biography, 1940–.
Written by the staff of H. W. Wilson, Co. (not by the subjects themselves).
Current Biography Cumulated Index, 1940–70.
Dictionary of American Biography, 1928–36.
Referred to as *DAB*. Index and five supplements up to 1977. A new reprint—the original 20 volumes, supplements, and index—was published in 1954. No living person in *DAB*. *Concise Dictionary of American Biography*, 1954 Abridged version.
Dictionary of American Negro Biography, 1982.
Covers up to 1970. No living person.
Dictionary of National Biography, 1885–1901.
British. Originally 63 volumes, now 20. Supplements every decade. No living person in *DNB*.
Concise Dictionary of National Biography.
Abridged. Vol. I (to 1900), 1952.
Vol. II (to 1950), 1961.
Directory of American Scholars, 1982–.
Covers humanities in four vols: 1, History; 2, English, Speech, Drama; 3, Foreign Languages, Linguistics, Philology; 4, Philosophy, Religion, Law—and Index.
Who's Who in America, 1899–.
Standard biographies, information often supplied by the subjects themselves.
Who's Who in the East (etc.).
Regional issues of *Who's Who in America*: East, West, Mid-West, South, and Southwest.

Who's Who in Art (etc.).
Issues of *Who's Who in America* by occupation: American Education, Commerce and Industry, etc.
Who's Who, 1849–.
British.
Who Was Who.
Poor's Register of Directors and Executives.
Guide to business executives, from Standard and Poor's, Corp.

Current Biographies of Men and Women in Science is a good source, but in some cases it may be easier just to use the directory of the relevant professional association. In the ape-language research, Duane Rumbaugh can be found in the APA directory.

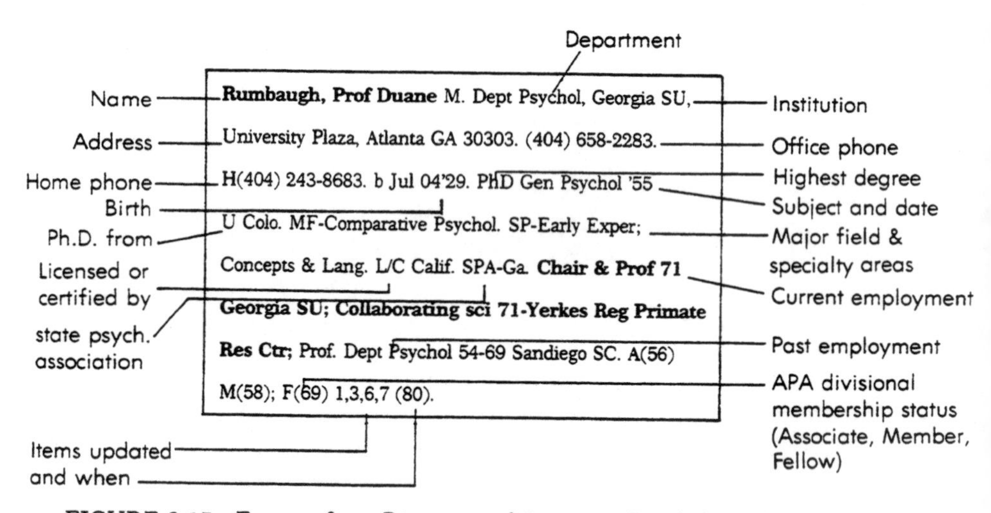

FIGURE 2.15 Excerpt from *Directory of American Psychological Association*, 1985 Edition

The abbreviations and symbols used are explained in the front of the directory. After Professor Rumbaugh's name you will find his address and telephone number(s), date of birth, highest degree and where it was earned, major field, specialty areas (up to two entries), licensure or certification as a psychologist (L/C Calif.), state psychological association membership (SPA-Ga.), principal current employment (bold type), other current or past employment, APA and divisional membership status (Associate, Member, Fellow), and directory recency symbol (items 1, 3, 6, 7 updated 1980).

ACTIVITY 15 Using *Current Biographies of Men and Women in Science*, *Directory of American Scholars*, or other biographical sources, find biographical information on one of the authorities in your research.

Book Reviews

Check your sources. Book reviews will help you evaluate the material you find. Many newspapers and magazines carry book reviews, so be sure to check the *New York Times Index* and The Magazine Index under the title or author that interests you. But there are also special reference works devoted to book reviews. For example:

Book Review Digest, 1905–.
Book Review Index, 1965–.
Index to Book Reviews in the Humanities, 1960–.
Technical Book Review Index, 1917–.

Linden, Eugene - *Silent Partners*

BL - v82 - Ap 15 '86 - p1164

Choice - v24 - S '86 - p159

KR - v54 - Ap 15 '86 - p615

LJ - v111 - Mr 15 '86 - p74

NYTBR - v91 - My 25 - '86 - p3

PT - v20 - My '86 - p80

PW - v229 - Mr 7 '86 - p88

SB - v22 - S '86 - p29

FIGURE 2.16 Excerpt from *Book Review Index, 1986 Cumulation*

Symbols are explained in the front of the Index. For example, NYTBR stands for *New York Times Book Review*; PT stands for *Psychology Today.*

Excerpt from Ursula K. Le Guin review of *Silent Partners* in *New York Times Book Review*:

> In his reporting of the vituperation, the increasing defensiveness of experimenters and the bad faith of some critics, the sensationalism and sentimentalism of much media coverage, Mr. Linden's candor and fair-mindedness contrast wonderfully with the prejudice and paranoia he describes. He does not pretend, however, to stand above the battle; his concern is ethical and urgent. For, after all the claims and counterclaims, now that we no longer see photographs of long, thumbless, inhuman hands signing "friend" in the language of the deaf, or read dismissive pronouncements from the lofty terraces of academic theory, now in the mid-1980's, what has become of the experimenters and their partners—Washoe, Lucy, Nim Chimpsky, Ally, Koko?

Reference Guides

There are many valuable sources to information in the library, more than can be illustrated here. Guides to the reference materials are available in the library itself:

Guide to Reference Books, Eugene Sheehy.
The New York Times Guide to Reference Materials, Mona McCormick.
Reference Books: A Brief Guide, Enoch Pratt.

The more you learn about your library, the more you will become a skilled independent researcher. This is one of the rewards of a college education, a useful skill you can take with you when you leave college. You need to develop the habit of spending time in the library, especially during those hours when the library is relatively empty. Learning how to use the library should be your first priority, your first step in becoming a researcher.

ACTIVITY 16 Using all the strategies in this chapter and as many different sources as you can find, collect a master bibliography on your research topic. Try to be as thorough and comprehensive as possible. Find newspaper, magazine, and journal articles; books; government publications—anything and everything that might have a bearing on your research. Look especially for the most recent sources.

STUDY GUIDE:
Chapter Two Using the Library

1. What is the "context" of a research question?
2. What is a search strategy?
3. What is the function of general encyclopedias in the search strategy?
4. What is a specialized encyclopedia? Does your library have any specialized encyclopedias in your area of research?
5. How many sources should be listed in your master bibliography?
6. How can you find a published bibliography on your research topic?
7. What is the function of the *Library of Congress: Subject Headings*?
8. In the excerpt from the *Library of Congress: Subject Headings* (Figure 2.5):
   a. What do "x" and "xx" indicate?
   b. What does "sa" indicate?
9. What information is given on a catalog card?
10. How tall is a book of 24 centimeters?

11. What is a full bibliographic reference?
12. What is a call number?
13. What is a "popular" source?
14. How could you find an essay (instead of a report) in an index?
15. Which is the most generalized index? How do you use it?
16. How can you find out what the symbols and abbreviations mean in an index?
17. In the excerpt from the *Readers' Guide* (Figure 2.9):
    a. Which entries indicate pictures present?
    b. Under "MEDOFF," what does 50+ indicate? How should this appear in your bibliography?
    c. Which entries indicate bibliographies present?
    d. What is the explanation of each of the items in the "MEGALITHIC" entry?
    e. Which entry indicates a coauthor?
18. True or false: The *Readers' Guide* is a good index for technical journals.
19. What are bound volumes of periodicals?
20. Why is it a good idea to limit yourself to some small aspect of a subject?
21. What is the general rule about where to start your research? Why?
22. What is a specialized index?
23. What is the new name for the *International Index*?
24. In the *New York Times Index* (Figure 2.10):
    a. Under BOMBS, which entries are cross-references; which are direct references?
    b. Under BOND, Beaulah, what does (S) indicate?
    c. Under BOOK Reviews, what is the meaning of each item under "Bailey"?
25. How can a library computer terminal be of use to a student?
26. What do LC and ISBN stand for?
27. What is a database?
28. What is ERIC?
29. How can you find out what's in ERIC?
30. Of what use might ERIC be to students?
31. What is in The Magazine Index?
32. How can you find a film review in The Magazine Index?
33. In what way is The Magazine Index superior to the *Readers' Guide*?
34. What is "Hot Topics"?
35. What is The National Newspaper Index?
36. What is The Business Index?

37. What is InfoTrac?
38. What is the meaning of "microform"?
39. What is microfiche?
40. Can you get a photocopy of an article on microform?
41. What is the *Congressional Directory*?
42. What is the main index for government documents?
43. Why might a researcher find book reviews like those in *Book Review Index* useful?

CHAPTER THREE

Reading Research

Reading is the foundation of research. Researchers must continually read professional journals, reports, and books in their field just to keep up with the endless stream of new studies and ideas. You must read to learn the background material in your own area of research. Although so much reading makes a lot of work for you, it also helps to make you expert in your field. You can use reading to educate yourself in the areas that interest you.

Researchers read for information; enjoyment is usually only a lucky secondary consideration. An exception to this may be found in research in the humanities. If you are reading literature, you should first read for enjoyment, and one type of research explores the sources of enjoyment in literature. But on the whole, researchers don't expect to be entertained by source material. They seek out the main points and subpoints and the organization of information; often they outline and take notes as they read. Their purpose in reading this way is to find out exactly what the author said. It's a deliberate pursuit of information—using reading to educate yourself, to teach yourself what you need to know in order to write.

Good researchers must be willing to go where the research takes them. You must not impose artificial limits on what you read. If you are researching a question about gun laws, you must not tell yourself, "I will read only the most recent articles that have to do with gun laws." The research itself will tell you what you must read. You may discover that you need to read the Constitution as background for an article on gun control—what does the Constitution say about your "right" to have a gun? You may also need to read articles on the *interpretation* of the Constitution. You will probably have to read books about gun control and so on. You will discover that the National Rifle Association has an interest in this question, so you must find out what you can about the NRA. And you will discover that there are organizations actively lobbying for new gun laws; you must find out who these people are, what their motives are.

As a skillful researcher, you must go below the surface of both style and message. If an author uses an unusual word or sentence, you as fellow writer must ask why—why this word here? Why this wording of the sentence? What is the author's purpose? What assumptions lie below the information? Where is this going: what implications follow from these ideas? (See Chapter 4.)

PURPOSE IN READING

Purpose controls reading: how you read depends on why you read. Whether you read fast or slow, carefully or casually, taking notes or not depends on your purpose in reading. To read research, you must have the ability to shift gears—to read faster or slower, for example—to suit different purposes in reading. It's important to understand that researchers don't read everything in the same way. If your only technique is slow, careful reading, you will soon find research tiring and very time-consuming. There are four general reading purposes in research: reading for facts, reading for analysis, reading for interpretation, and reading for evaluation (critical reading).

Reading for Facts

Researchers read for facts, information. We read not only to find out what the facts are, but to be able to recall them later, to arrange them in new patterns that have meaning for us. A researcher's aim is to be able, eventually, to work without notes. Researchers must not only find but assimilate facts. To assimilate information means to internalize it so you can recall it, manipulate it, work with it; it means to make the information part of your own working knowledge. Thus, reading for facts is really a way to teach yourself; the more knowledge you acquire in this way, the more expert you will become.

Researchers don't really read, in the sense of reading a novel, when looking for facts. This kind of reading is more like studying. It's the kind of reading most people do when they use dictionaries, encyclopedias, almanacs, and so on. To write about Napoleon, you need to know as much as possible about the man, about France and French history. Fact finding calls for gathering up all the books and articles you can find on your subject, reading quickly, isolating the facts, taking notes.

ACTIVITY 17 Read the following excerpt. What is its main point? (How do you know?) What facts are presented? Which facts are worth noting for research? (How do you know?)

> Can a kitten raised by a dog find happiness with a 230-pound gorilla who converses in American Sign Language? Well, you can ask Koko, a 13-year-old

female lowland gorilla who for a dozen years has been the focus of the world's longest ongoing ape language study, sponsored by the Gorilla Foundation of California with past support from the National Geographic Society. According to Dr. Francine "Penny" Patterson, Koko uses more than 500 signs regularly and knows some 500 others in American Sign Language, or Ameslan—the hand language of the deaf. That's how Koko came to tell Penny that she wanted a cat—a word she signs by pulling two fingers across her cheeks in the manner of whiskers. So Penny gave her a toy cat. Koko pouted.

Then last June a litter of three kitttens was brought to the rural compound near San Francisco where Koko lives with Penny and Michael, an 11-year-old gorilla also versed in Ameslan. Abandoned at birth, the kittens had been wet-nursed by a cairn terrier for four and a half weeks. "Love that," signed Koko to the kittens. Gingerly examining them, she chose the tailless male and named him All Ball.

Jane Vessels, "Koko's Kitten," *National Geographic* 167.1 (1985): 110.

Reading for Analysis

We read to analyze many things: the sequence of events; the organization of the information; the validity of the reasoning; an author's assumptions, attitudes, biases. Analysis usually means taking apart, dividing into components, breaking down into smaller parts. Facts—isolated data—alone won't tell you everything you need to know. You need to know also how they are put together, how they are arranged. For example, you may have found many facts about Napoleon's virtues—his strength as a leader, his brilliance as a military general—but has the author also presented the other side—Napoleon's weaknesses, faults? What can you safely deduce from a one-sided presentation of facts?

Outlining for Analysis

One of the simplest and most effective methods of analyzing books and articles is outlining. You need to know the main points and subpoints of anything you read in research; you need to know how authors have organized their ideas. Nothing is more efficient for revealing the content and structure of a piece of writing than outlining. There are different sorts of outlines, depending on your purpose in outlining, especially who you are outlining for. In a formal outline you need to take care to see that you follow all the rules of outlining.

Outline Styles An outline must be clear and efficient; it's intended to convey maximum information in minimum space. You may choose either the sentence outline or the topic outline, whichever suits your purpose best, but avoid doing both in the same outline. Traditionally, parallel points are written in parallel language, and the numbering is aligned on the periods.

Parallel language:

Advantages of Jogging
- I. General health improves
- II. Physical energy increases
- III. Psychological outlook improves

Nonparallel language:

Advantages of Jogging
- I. Better health
- II. Energy increases
- III. Improvement of outlook psychologically

Sentence Outline

Smoking Should Be Banned
- I. Smoking is harmful.
- II. Smokers suffer coronaries and lung cancer.
- III. Nonsmokers suffer from passive smoking.
- IV. Everyone will benefit from a smoking ban.

Topic Outline

Smoking Should Be Banned
- I. Hazards of smoking
- II. Dangers to smokers: coronaries and lung cancer
- III. Dangers to nonsmokers: passive smoking
- IV. Solution: smoking ban

Note that in formal outlines every *I* must have a *II*, every *A* must have a *B*. A category of subdivision containing only one item doesn't make sense. For example:

- I. Cause of pellagra
  - A. Diet deficiency of niacin and protein

If you find yourself with such an item on your outline, you should reword the item so that the subpoint becomes part of the main point:

- I. Cause of pellagra: diet deficiency in niacin and protein

Occasionally it may be possible to reword so that there is a new main point with two subpoints:

- I. Diet deficiencies causing pellagra
  - A. Protein
  - B. Niacin

The most probable solution to this problem is more research; more data are probably needed on that particular point in your outline:

I. Causes of pellagra
   A. Diet deficiencies
      1. Niacin
      2. Protein
   B. Alcoholism

The progression in a formal outline moves from roman numerals to capital letters to arabic numerals to lowercase letters to arabic numerals in parentheses to lowercase letters in parentheses:

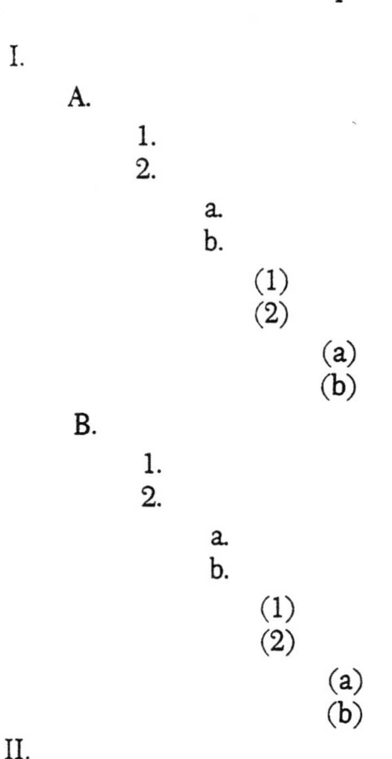

I.
   A.
      1.
      2.
         a.
         b.
            (1)
            (2)
               (a)
               (b)
   B.
      1.
      2.
         a.
         b.
            (1)
            (2)
               (a)
               (b)
II.

- *Additional Subdivisions*. The subdivisions here should be adequate for any outline in college. Theoretically there could be more levels or subdivisions, but the need for such fine divisions is rare.
- Some possibilities include using lowercase roman numerals (i, ii, iii, iv), double letters (aa, bb, cc), or brackets [I], [A].
- There is no standard procedure for additional subdivisions. (See the decimal outline, below.)

- *Capitalization.* Capitalize the first word and *only* the first word of all lines of the outline.
- Occasionally the rules of capitalization may require otherwise, especially if there are proper nouns in your outline: I. Twelve major forces in Poland.
- Use periods after all numbers and letters except those in parentheses; numbers and letters should be aligned on the periods.

Aligned on periods	*Not Aligned*
I. History of the problem	I. History of the problem
II. Suggestions from the left	II. Suggestions from the left
III. Suggestions from the right	III. Suggestions from the right

For very highly detailed outlines with many subdivisions, consider using the decimal style:

1.0 First main point
    1.1 First subpoint
    1.2 Second subpoint
    1.3 Third subpoint (and so on)
        1.3.1 First subpoint under subpoint 3
        1.3.2 Second subpoint under subpoint 3
            1.3.2.1 First point under preceding point
            1.3.2.2 Second point under 1.3.2
2.0 Second main point

The decimal outline is capable of infinite division and therefore may be useful in science, some aspects of government and business, and law. Since it's a specialized form of outlining, check with your instructor before using it in college.

Read the following excerpt. How would you outline it?

> It was hard to tell how Humphrey the whale was holding up last week, his third spent cruising the delta waters he was not supposed to survive for more than about 14 days. Some onlookers said the humpback's skin had turned from black to an ominous gray in the fresh water of California's Sacramento River (though you couldn't always hear them above the sound of Humphrey's tail slapping the surface in a display of healthy anger). A brain parasite might have caused the 40-foot, 40-ton whale to veer from its Pacific migration path, take a side trip under the Golden Gate Bridge and up into one of the two rivers that spill into San Francisco Bay, some scientists believed. But the press gave more play to the possibility that Humphrey was actually a pregnant female coming into warmer shallow waters to calve, a theory proffered by one state bureaucrat. Such political people have been out in force along the winding Sacramento, exerting pressure

on the assembled scientists to quit letting nature take its course and, for God's sake, *do* something.

"Trying to Save the Whale," *Newsweek*, November 11, 1985, p. 39.

Outline of Excerpt from Article Note that the outline condenses and interprets; it isn't made merely by lifting sentences from the original:

Concern for Humphrey the Whale

I. Humphrey survives three weeks in delta (fresh) water.
   A. Whales survive only 14 days in fresh water.
   B. Humphrey turns color (but seems healthy).
II. Theories offered about why whale entered river.
   A. Brain parasite disorients whale.
   B. Pregnant whale seeks warm shallow water to calve.
III. Pressure mounts for scientists to do something.

The outline shows major and minor points. Your outline can have as much or as little detail as you need. A "complete" outline will account for everything in a composition, but for study purposes you may not need so much detail. If the outline is being made for someone else (like an instructor), you must follow the conventions of outlining carefully.

ACTIVITY 18 Outline the following passage (a continuation of the whale story above). Assume you are outlining for someone else—an instructor or an employer—use formal outlining conventions.

> And so, as the whale wandered willfully in and out of sloughs and tiny tributaries, things got done. More than 200 volunteers and government workers, in fact, banged on pipes, played tapes of killer marine mammals and tried to herd Humphrey with their boats. Psychics paddled in his wake, thinking they could direct the whale with brain waves. The idea was to get Humphrey back at least to the saline waters of the bay. But by the weekend rescuers had failed to even affix to him a radio transmitter that might save hours spent just trying to find him each morning. Humphrey moved too quickly and purposefully, albeit never for very long in the same direction.

Rhetorical Analysis

Another significant way to look at information is called *rhetorical analysis*. Rhetoric is a very powerful analytical tool for many kinds of writing. In any situation in which people use language—oral or written—to communicate, there are always three main components: the speaker (or writer), the subject matter, and the audience (or reader). Each of these components will yield significant insights into the written material, and if you add to them your analysis of the writer's purpose, you will produce a thorough analysis.

The Writer's Purpose Why has the author written the report, essay, book? Is the author reporting research? criticizing research? Purpose governs everything in writing, and you must determine whether you are reading objective research, an opinion, or something else. Think about the difference between articles in professional journals and articles in "popular" magazines; writers' purposes are influenced by their audiences.

You must look both at what the writer says and how he or she says it to determine the writer's purpose. Frequently the author's purpose involves not only what is said but also what is left unsaid. You must determine the author's assumptions, unspoken or implied premises—what is *hidden* from the uncritical reader. Read carefully to discriminate between an author's *apparent* purpose versus the *real* purpose.

1

THREE MILE ISLAND

The high and growing cost of nuclear power plants is due not so much to the difficulties associated with the technology that it has in common with non-nuclear plants—that is, the conversion of energy of steam into electricity—but rather to its unique feature, the use of fission to supply the heat needed to produce steam. The accident at Harrisburg showed that a failure in the steam-to-electricity section of the plant that would have caused very little trouble in a conventional power plant came close to producing a catastrophic disaster in the nuclear one and has shut down the plant for a long time, and possibly permanently.

High costs related to use of fission

Accident caused by non-nuclear conversion of steam

The Three Mile Island Power Plant produced the steam needed to drive its electric turbines in a pressurized-water reactor. In such a reactor, water is circulated through the reactor's fuel core where—because it is under pressure—it is heated far above its normal boiling point by the heat generated by the fission reaction. The superheated water flows through the reactor's ''primary loop'' into a heat exchanger where it brings water, which circulates in a ''secondary loop,'' to the boiling point, and the resulting steam flows into the turbine to generate electricity. The spent steam is recondensed and pumped back to the heat exchanger, where it is again converted to steam, and so on. The third loop of cooling water is used to condense the steam, carrying off the excess heat to a cooling tower where it is finally released into the air. This arrangement is much more complex than the design of a conventional power system, where the steam generated in the boiler passes directly into the turbine. In this type of nuclear plant the water that circulates through the reactor (which is equivalent to the boiler in a conventional plant) becomes intensely radioactive, and the complex successive circulation loops are essential to keep that radioactivity from leaving the reactor.

Fission produces hot water

Hot water flows to heat exchanger, which produces steam

Cool water condenses steam; heat is released into air

Circulating water becomes radioactive

Pump fails, water overheats

Release valve sticks open, too much water is lost

Intense heat threatens meltdown

Emergency cooling system activated but turned off too soon

Hydrogen bubble blocks flow of cooling water, meltdown again possible

Explosion of gas possible

Technicians reduce danger of explosion

Plant remains highly radioactive

On March 28, 1979, at 3:53 A.M., a pump at the Harrisburg plant failed. Because the pump failed, the reactor's heat was not drawn off in the heat exchanger and the very hot water in the primary loop overheated. The pressure in the loop increased, opening a release valve that was supposed to counteract such an event. But the valve stuck open and the primary loop system lost so much water (which ended up as a highly radioactive pool, six feet deep, on the floor of the reactor building) that it was unable to carry off all the heat generated within the reactor core. Under these circumstances, the intense heat held within the reactor could, in theory, melt its fuel rods, and the resulting ''meltdown'' could then carry a hugely radioactive mass through the floor of the reactor. The reactor's emergency cooling system, which is designed to prevent this disaster, was then automatically activated; but when it was, apparently, turned off too soon, some of the fuel rods overheated. This produced a bubble of hydrogen gas at the top of the reactor. (The hydrogen is dissolved in the water in order to react with oxygen that is produced when the intense reactor radiation splits water molecules into their atomic constituents. When heated, the dissolved hydrogen bubbles out of the solution.) This bubble blocked the flow of cooling water so that despite the action of the emergency cooling system the reactor core was again in danger of melting down. Another danger was that the gas might contain enough oxygen to cause an explosion that could rupture the huge containers that surround the reactor and release a deadly cloud of radioactive material into the surrounding countryside. Working desperately, technicians were able to gradually reduce the size of the gas bubble using a special apparatus brought in from the atomic laboratory at Oak Ridge, Tennessee, and the danger of a catastrophic release of radioactive materials subsided. But the sealed off plant was now so radioactive that no one could enter it for many months—or, according to some observers, for years—without being exposed to a lethal dose of radiation.

Radioactive gasses escape from plant

Some radioactive gases did escape from the plant, prompting the Governor of Pennsylvania, Richard Thornburgh, to ask that pregnant women and children leave the area five miles around the plant. Many other people decided to leave as well, and within a week 60,000 or more residents had left the area, drawing money from their banks and leaving state officials and a local hospital shorthanded.

Nuclear plants vulnerable to catastrophe

Like the horseshoe nail that lost a kingdom, the failure of a pump at the Three Mile Island nuclear power plant may have lost the entire industry. It dramatized the vulnerability of the complex system that is embodied in the elaborate technology of nuclear power. In that design, the normally benign and easily controlled process of producing steam to drive an electronic generator turned into a trigger for a radioactive catastrophe.

Barry Commoner, The Politics of Energy, Knopf, 1979

The author's point of view is primarily one of objective reporting. He presents facts: "But the sealed-off plant was now so radioactive that no one could enter it for many months—or, according to some observers, for years—without being exposed to a lethal dose of radiation." Yet there is a certain amount of emotional language—he uses *disaster* and *catastrophe* several times, despite the fact that no one was injured and there was no damage outside the plant itself. His final sentence relies on a metaphor (the plant turned into a "trigger for a radioactive catastrophe"). Thus, although he seems to be objectively reporting facts, he is coloring the presentation—putting it in the worst light and attempting to frighten the reader. His real purpose is argumentative: he is trying to persuade the reader that nuclear power is dangerous.

The Writer Who is the writer? You can often find out who the author is in one of the *Who's Who* volumes or other sources. Credentials don't mean everything in research, but it's helpful to know whether an author is an authority, a recognized expert in the field or a relatively unknown researcher. More important than writers' credentials are their *voices*, the way they project themselves through their writing. Writers who sound dull, flat, or boring seriously prejudice their own writing, as do writers who sound too peppy, enthusiastic, or more interested in making an impression on the reader than in the research. For example, listen to the sarcasm in this writer's voice:

> In his book, Morganthaler makes the incredibly stupid claim that vultures have no olfactory receptors. It's obvious that Morganthaler doesn't know what he is talking about. Give me a break. Does Morganthaler think vultures can *see* dead animals twenty miles away? Has Morganthaler ever heard of a turkey vulture? Read Morganthaler if you're interested in cactus or boat building, but not vultures.

It's possible to maintain an appropriate balance between a heavy, pedantic voice; a sarcastic voice; or a frothy, ingratiating voice. The proper "voice" for most research writing is businesslike and professional.

Of greatest importance is the writer's *credibility*. Research has little value if we can't trust the researcher. We expect the research to be accurate, the writing to be clear and concise. Any suggestion that the writer may be exaggerating, taking shortcuts, making mistakes, using faulty reasoning, and the like will alienate readers.

- Aristotle said that the most important factor in rhetoric was the character of the speaker. The Greek word for character is *ethos*, and Aristotle said that appeals to ethos were even more important than logic and evidence. People can't always detect faulty logic, nor can they always verify evidence, but everyone can form a judgment about a writer's character.
- We analyze writers' characters by what they say and how they say it. Written words are clues to character, and that includes "invisible" clues

like attitudes, assumptions, and even what the author doesn't say. When a writer omits something we think should have been included, fails to answer a question we think important, avoids or misses research we think relevant—most people will begin to form a negative opinion about the writer.

- To analyze the writer, we must ask not only who the writer is, but who the writer is pretending to be: What role or personality is the writer projecting? Who does the writer *think* he or she is? What authority does this writer have for writing on this subject?

ACTIVITY 19 Read the following excerpt. The author has written many essays for a monthly column in *Natural History* magazine. What can you tell about this author's voice? his personality? Discuss the author's ethos.

> Throughout a long decade of essays I have never, and for definite reasons, written about the biological subject closest to me. Yet for this, my hundredth effort, I ask your indulgence and foist upon you the Bahamian land snail *Cerion*, mainstay of my own personal research and fieldwork. I love *Cerion* with all my heart and intellect but have consciously avoided it in this forum because the line between general interest and personal passion cannot be drawn from a perspective of total immersion—the image of doting parents driving friends and neighbors to somnolent distraction with family movies comes too easily to mind. These essays must follow two unbreakable rules: I never lie to you, and I strive mightily not to bore you. But, for this one time in a hundred, I will risk the second for personal pleasure alone.
>
> Stephen Jay Gould, *The Flamingo's Smile: Reflections in Natural History*, New York: Norton, 1985: 197.

The Subject Our analysis of subject matter depends a great deal on our own expertise. For that reason, you must read everything you can find on the subject, make yourself as expert as you can through self-help, reading and studying all the available research. You must be able to judge the accuracy of the research, a difficult job unless you have read many books and articles on the subject. You need to know something about research design. When researchers talk about "probability," they assume other researchers will know what they are talking about.

But mostly we require that the composition, book, essay, or report make sense. We judge subject matter on logical criteria. The Greek word for logic is *logos*, meaning "the appeal to reason." Through *logos* the writer offers us information, data we can think about, reason with. Whereas *ethos* appeals to our sense of morality and ethics, and *pathos* appeals to our emotions, *logos* appeals to the mind. Thus we analyze subject matter on "reasonable" grounds. We ask how much information the researcher has amassed (is there enough?), how well the information has been organized for clear and concise communication. And we are especially interested in the researcher's assumptions, inferences, and conclusions. If we detect any logical fallacies, any bias in the data or procedures, any faulty conclusions (that do not follow from the researcher's premises), we conclude that the research isn't good—it isn't accurate, which is tantamount to saying it isn't research. And of course we will conclude that the researcher isn't competent—or worse, not honest.

At the very least we require the researcher to express an appropriate attitude toward the subject. A writer who somehow manages to treat cancer as a joke, or who seems not to respect the subject matter, will certainly alienate most readers. On the other hand, researchers who seem excessively impressed with their own research usually sound pompous, pedantic, full of importance. When we say writing has a heavy "tone," we mean the writer is taking it much too seriously, treating the subject with far more reverence and importance than we think is appropriate:

> The function of modern political advisors is to hyposensitize the electorate to candidate-specific attributes and issue-related factors so that in a uniformity of standardization, candidates appear isometric, thereby depriving the electorate of any meaningful decision-making discriminants other than those manufactured by the advisors.

When writing is this ponderous and unnecessarily technical, most readers feel there is something wrong with the writer. That is, such writing is seldom perceived as necessary technical jargon; it sounds more like *unnecessary* jargon—the writer is trying to make a nontechnical subject sound like science. In this case, bad writing reveals the writer's insecurity. The writer tries to impress the reader but winds up sounding like a kid using every big word in the dictionary.

ACTIVITY 20 Read the following excerpt. The author is describing nuclear physicists' attempt to understand the way barium might be split off from a nucleus of uranium. What is the author's attitude toward this subject? How does the author's use of *logos* add to the credibility of his writing?

> They pictured the uranium nucleus as a liquid drop gone wobbly with the looseness of its confinement and imagined it hit by even a barely energetic slow neutron. The neutron would add its energy to the whole. The nucleus would oscillate. In one of its many random modes of oscillation it might elongate. Since the strong force operates only over extremely short distances, the electric force repelling the two bulbs of an elongated drop would gain advantage. The two bulbs would push farther apart. A waist would form between them. The strong force would begin to regain the advantage within each of the two bulbs. It would work like surface tension to pull them into spheres. The electric repulsion would work at the same time to push the two separating spheres even farther apart.
>
> Richard Rhodes, *The Making of the Atomic Bomb*, New York: Simon, 1986: 259.

The Reader Much research is aimed at experts. Researchers write for other researchers, and this can make it hard on the rest of us. When experts write to experts, they take shortcuts, use specialized jargon, and in general write at a level that nonexperts find hard to follow.

> Formally we can say that a formative must be regarded as a pair of sets of features, one member consisting of the "inherent" features of the lexical entry or the sentence position, the other member consisting of the "noninherent" features introduced by transformation. The general principle for erasure operations, then, is this: *a term* X *of the proper analysis can be used to erase a term of the proper analysis just in case the inherent part of the formative* X *is not distinct from the inherent part of the formative* Y. But notice that this is an entirely natural decision to reach.
>
> Noam Chomsky, *Aspects of the Theory of Syntax*

Most of us find this passage by the linguist Chomsky difficult to read (he is discussing the theoretical problems of a new grammar). In part it's hard to read because it is technical subject matter: to understand it we must know what is meant by *formatives, features* (inherent and noninherent), *erasure operations*, and so on. Secondly, Chomsky is writing to other linguists, and for that reason he has chosen to use technical language—it's accurate and a kind of professional shorthand. Writers' views of their readers can have a significant influence on research writing. When writers assume the reader is an expert, they are likely to take less care to see that their writing is readable. On the other hand, some writers can sound very condescending when writing for the general public. So-called "popularized" research can resort to cuteness or focus on the spectacular and dramatic aspects of research instead of on the technical presentation of data.

Appeals to the reader, especially those designed to arouse his or her curiosity or approval, are frequently based on emotion. The writer tries to excite the reader, amuse the reader, or sometimes frighten or stir up the reader. Appeals to emotion Aristotle called appeals to *pathos* (the origin of our words *sympathy, empathy,* and *pathetic*). Many researchers feel that pathos has no place in research writing. Objective, logical presentation of data, they feel, is the only appropriate procedure for research. However, such a judgment is too limited: everything depends on purpose. In the humanities, for example, pathos is a legitimate and serious element of research. In the sciences, on the other hand, appeals to emotion have little place in research, yet, surprisingly, researchers aren't always aware of how much their own emotions distort their research. As skillful readers of research, we must all be alert to inappropriate intrusions of emotion into research reports. If you read articles from the popular press, you are almost certain to find writers trying to make their readers feel sorrow, pity, anger, and other emotions. Surprisingly, even scholarly journals aren't entirely free from emotionalism. (Almost all researchers working with apes in the ape-language research, for example, give the apes names and begin to treat them as if they were human, including kissing and hugging them.)

Attitudes toward the reader can make writing seem either far above the reader's head—too difficult to read or understand—or far below the reader's intelligence. Somewhere between these two extremes is an appropriate level for the general, educated reader. The writer constructs a model or image of the reader, and writes to this hypothetical reader. Ask yourself, "Who is the reader?" That is, who does the writer think the reader is? What is the writer's image of the reader? An expert? A general reader? You must be especially alert for writers who appear to be writing for one audience but in fact are aiming their message at another. For example, in a court case, a clever lawyer asks questions of the witness so that the jury will hear the answers. And sometimes in his speeches to the nation, the President may appear to address his remarks to Congress, but of course the rest of the nation hears these remarks as well.

ACTIVITY 21 Read the following student essay. Look carefully at what the author says and how she says it. Analyze the author's rhetoric: purpose, ethos, pathos, and logos.

1

LICENSE TO KILL—USED OR ABUSED?

The deer returns to the same spot every day. Just at the edge of the woods, in a clearing, awaits a small clump of hay, and daily he returns to feed on it. Today the deer steps carefully through the underbrush, the crunch of fallen autumn leaves underfoot. He approaches the hay cautiously, as usual, and begins to feed. Suddenly, in the quiet of the November morning, a shot rings out—and the deer lies dead in a pile of blood-soaked hay. It is the opening day of hunting season.

This is just one example of the things that go on every year in my neighborhood-and countless other rural areas; and it sickens me. I have grown up amidst deer in the wild, and I know what beautiful living things they are. Yet every year I see my neighbors and my friends slaughtering these lovely animals. I do not advocate a ban on hunting; on the contrary, I believe that hunting must be permitted to mercifully prevent the starving of an overpopulated herd. However, hunting is a privilege, and the illegal abuse of this privilege is becoming a common practice—far too common to be ignored.

The reasons for hunting have been blown completely out of proportion. It started out as a means of survival—to capture food necessary for sustenance. As it became less and less a necessity to hunt for food, hunting became more and more a sport. It was a leisure-time pursuit enjoyed by the men—it was a ''man's sport'' in every sense of the term. And, of course, it eventually became a proving ground for one's manhood—''the great lion hunter'' symbolizing man's conquest and masculinity. This, I think, was the turning point from a ''fair game'' sport to a highly competitive contest. Simply getting a deer was not enough; a man's skill and ability with a gun was measured by the size of the deer or the points on a buck's antlers. Such ''trophy'' bucks lead to the hunters being choosy about the deer they shoot—they ultimately try for the biggest, strongest buck in the herd.

Thus with its competitive nature, hunting became a testing ground for man's cunning in how to outsmart the deer, so that the man could add another set of antlers to his collection. And here is where the abuse of hunting came into being.

Each state issues a certain number of buck-hunting licenses and a fewer number of doe permits. Once the hunter has his hunting permit, he is free to shoot but only during the hunting season—November 15 to November 30. He is also limited in hunting only during the daylight hours. Each license allows its holder to kill only one deer, and once the hunter has killed one animal, he may not so much as shoot at another. Also if a hunter should shoot a deer but fail to kill it, he must track down the deer until it dies or until the hunter can shoot it again. With these few simple rules, the oversupply of deer can be eliminated and the deer population can be kept under control. But this is hardly what happens each year as the sportsmen take to the woods, guns in hands, to capture their prizes.

I live in what I consider to be an excellent cross-section of deer-hunting areas in our country. It is rural, middle-class, mid-America, and the annual hunting season is a popular event. The first in our neighborhood to get a deer is hailed as something of a celebrity, so obviously everyone is out to beat the other guy in getting his deer. The contest has begun.

Tim, 17, a high school senior, on his first day out hunting, spots a herd of about ten deer, all running away and out of range. Suddenly a lone animal comes limping out of the woods, frantically trying to catch the rest of the herd. Tim sees movement, turns, and fires, only to catch the poor, small, sickly doe in a rear leg. After realizing the condition of the pitiful animal, Tim turns and walks away. Who needs a small yearling like that?—no antlers anyway. And in the distance the doe limps on in pain.

A mile away, Howard Mailer and his son Bob have

been out all day, without sight of a single deer. It's getting darker and darker outside—almost supper time. On their way back to the truck, they spot a set of prints, and another, and another, until a well-defined trail is obvious. Wait until tomorrow? Not a chance—the tracks are still fresh. Bob runs back to the truck and gets the searchlight. The two men follow the tracks—the searchlight beaming ahead into the darkness. Minutes later, the lamp lights upon two tiny, bright specks. Immobilized, the two eyes stare transfixed into the beam. Howard pulls up his rifle and shoots. The eyes disappear and a body heaps to the ground. ''You got him, Dad!'' Bob announces happily and proceeds to slash the deer's throat.

Bill Ozwald, another neighbor, has his own sure-fire method. Every day Bill has taken hay to a secluded, quiet niche, and every day the hay has been eaten by a small group of deer. Today Bill will kill one of those deer.

These are just three of the ingenious examples my neighbors have been able to come up with ''for the sport of it all''—and I'm sure there are many more of these not only illegal but immoral practices going on. Not only are they breaking the law of the state, but they are also breaking the laws of nature. Laws of predation allow for the weak and slow to get caught and killed. This keeps the population down and also allows for survival of the fittest. On the contrary, modern day hunters many times, with the combination of high-powered, efficient rifles and illegal hunting techniques, weed out and kill the cream of the crop—the strongest buck with the most points on his antlers. This upsets the balance of nature, for the continuation of a species depends upon its ability to survive and procreate the most fit of its kind.

I'm not saying that all hunters hunt like this. But as I have observed, and by the hearsay of my neighbors, I know these types of hunting strategies are practiced. It sickens me to think how low a person will stoop to prove he can kill. And

I do not think that killing a deer makes a man big; it takes a pretty small person to have to break the law to prove something about his masculinity. Hunting is a sport—and it should be a fair one. An impossibility? Maybe. But maybe, too, if people are made aware of what's going on in the great outdoors, the importance placed on hunting and killing will not be so great. At least with greater public awareness we should be able to stop the abuses of hunting.

Laurie Stewart

Reading for Interpretation

Most researchers hope that their work won't need to be interpreted. We all wish we could say our essay or report means just what it says, especially in the sciences (literature, philosophy, religion, and so on are another matter), that there is a one-to-one correspondence between what it says and what it means. Unfortunately, very little research is written with such clarity, such simplicity of interpretation. Even if we have understood perfectly every word in a text, we may still need to ask, so what? What is the point? What does it *mean*?

We read to interpret ideas, the meaning behind the words. Some words are more difficult to interpret than others; they have literal and nonliteral meanings. "The sky is gray." Literally this is a description of the sky, its color. But we attach a connotation to "gray sky": it's going to rain, the day is dreary, and so forth. Some kinds of research (in the humanities, for example) may deal with symbolism and metaphoric language. Much research is written in technical language—the special jargon of chemistry, physics, sociology, and so on—making the job of the student researcher especially difficult. Interpretation can be influenced by complex, unusual, sophisticated sentences as well as difficult vocabulary. While many readers are interested only in facts and information, more critical readers must remain alert to the effect of style and attitude on meaning. Not all research is written in the simple, unambiguous language of a lab report.

We must read for meaning if we want to interpret an author's purpose, intent, mood, and tone. They are deductions on our part. Interpretive reading allows us to draw inferences, predict the unknown from the known. It also allows us to extrapolate—to extend the author's ideas; make comparisons, analogies; draw implications, conclusions. It is meaning that supplies the context for facts and analysis. A list of facts is simply a list until we know what they mean. An analysis of an event is simply an intellectual exercise until we can supply the appropriate context. Facts and analysis allow us to understand meaning; meaning illuminates facts and analysis.

ACTIVITY 22 Read the following excerpt. Read for facts and analysis and then discuss your understanding of its meaning.

> . . . Jarvis Bastian, in 1966, reported in a highly imaginative experiment. Two Atlantic bottlenose dolphins, a male and a female, were kept in a large tank and trained to work together in pressing paddles to be rewarded by an automat which disgorged fish when the proper paddles were pressed. At a certain stage in the complicated sequence of training procedures, the tank was divided by an opaque partition, with the male on one side and the female on the other. The arrangement was then as follows. Both the male and the female were warned by lamps being switched on that the game was ready to begin. Then another lamp was switched on to give *either* a continuous *or* a flashing light. In the former

case, the right-hand paddle must be depressed, in the latter case the left-hand paddle. Now the female could see this signal lamp, but the male *could not see either the lamp or her*. Both dolphins got fish if, and only if, the *male first* pressed the correct paddle on his side of the tank, and *then the female* pressed the correct paddle on her side, there being a pair of paddles for each of them. So the male had to press the correct paddle without seeing the lamp that signalled which paddle to press. On the face of it, he could only do this *if the female told him, by her calls, which paddle to press, when she saw whether the lamp was flashing or steady*. Nevertheless, over many thousands of runs, the male pressed the correct paddle and the dolphins succeeded in earning their reward on more than 90% of the tests. Analysis of the female's calls indicated that she made different pulsed sounds when her lamp was flashing and steady, responding sooner, longer and at a faster pulse rate to the steady light; it is quite possible the male, hearing her, could tell the difference between the two kinds of calls. The dolphins' success was only prevented if *either* the female was not rewarded with fish (as happened accidentally in two test series through a defect in her automat), *or* her signal light was hidden from her as well as from the male, *or* the barrier between male and female was made sound-proof. It seems certain from this amazing experiment that *in some sense the female was telling the male whether the light was flashing or steady, so that he could press the correct paddle in response*.

Claire Russell, and W. M. S. Russell, "Language and Animal Signals." *Language*, second edition. Ed. Virginia P. Clark et al. (New York: St. Martin's, 1977): 60.

Reading for Evaluation

We read to evaluate the quality of a work, the value of its information, the impact of the work on the research question, the philosophical orientation of the book, the skill of an author's style and techniques, the overall significance. Reading to evaluate is critical reading. Factual reading answers the question What does it say? Analytical reading answers the question How is it constructed? Interpretive reading answers the question What does it mean? And evaluative reading answers the question So what? What good is it? How important is it?

In order to read critically, we must first read factually (making sure we understand what the author has said), read analytically (making sure we understand how the essay, article, or report is structured and so on), and read interpretively (making sure we understand what the author means). Only then can we attempt to evaluate, to judge the worth of the writing. We owe it to every author to be as thorough in reading for facts, analysis, and interpretation as possible before reading critically. Researchers who get too critical too soon may bias their evaluation, distort through lack of understanding. Student researchers, especially, must take care to read carefully before reading critically. Much published criticism is refuted by authors who charge that the critic hasn't read carefully, hasn't understood what he or she was criticizing.

Critical Criteria

To evaluate we must go outside the text; to judge its significance we must supply a context—criteria against which to measure. It's difficult to specify a universal set of criteria by which we might evaluate anything and everything. In the sciences we may wish to answer the basic question, What is the truth? But in literature, music, and art we don't usually ask for truth. For the time being, then, let us limit ourselves to some general criteria for nonfiction writing.

Evaluating Content Content is the subject matter of a composition, what it's about. It is the information. If you write about the disease AIDS, whatever you say about AIDS is the "content" of your paper (as opposed to its style). Theoretically, the information in a piece of factual writing would be the same even if presented in a foreign language, sign language, or pictograms. This information is the "content"; it's the most obvious kind of evidence.

1. What is the information: What point does the author make? What are the facts in this composition? Is this a thorough treatment, a summary, or a sketchy overview? Is there enough data here? Has the author supplied enough facts or details to achieve his or her overall purpose? Has the author covered everything you think he or she should, in as much detail as you think the subject deserves?
2. What is the author's purpose? What kind of material is this? Opinion? Editorial? Argument? Factual report (news)? Is the writer objective (presents both sides fairly) or biased in favor of one side? Does the author present evidence to justify his/her stance? What is the author's intent: What is he/she trying to do? Can you detect any difference between the author's real purpose and apparent purpose?
3. Is the author accurate? Does the author have the facts straight? Or can you tell that he/she is distorting, exaggerating, or diminishing the facts? You must be well read in the subject before you can answer a question like this. You must be able to say that author X agrees with authors Y and Z, for example. General facts, like the national debt, can be checked against almanacs or other sources, but for more specialized information pertaining to specific research topics, you need a background in that subject.
4. Where did the author get the information? Experiments? Interviews? (Named sources or unnamed?) Investigative reporting? Books and documents? How reliable are these sources? If the author doesn't say where the information came from, how credible is the author?
5. What are the author's assumptions? What beliefs or attitudes must the author hold in order to write this article? (If the author believes what he or she is saying, what else must he or she believe?) What implications

follow from this article? If we believe what the author is saying, what else must we believe as a logical consequence? Because assumptions, beliefs, and so on are invisible (not visible on the printed page), this kind of analysis is very sophisticated. Often we cannot point to an author's assumptions; we must deduce them. (See also in Chapter 4, Evidence.)

6. Is the author's reasoning good? Does he or she give examples and illustrations to prove a point? Are cause-and-effect arguments believable? Do the author's conclusions follow from his or her premises? Can you detect errors or fallacies of logic?
7. Who is the author? What are his/her credentials? What right does this writer have to deal with this subject? How credible is this author? Where are the clues to the author's credibility? Does the author project a reliable, trustworthy personality? Where are the clues to the author's voice?
8. Is this subject new? Is it a new subject or a new development of an old subject? Is there any new information or new interpretation of facts here? What is the relation between this article and others written earlier on this subject—what is the history of this subject?
9. Is this a significant subject? Does it treat a subject most readers would agree is worth treating? If it seems trivial or light, did the author intend it that way? Is it interesting to you personally, and would it be interesting to the general, educated reader? Why is it significant—why is it worth researching, for example?
10. What is your overall judgment of the content of the article? Is the article good? Did you feel satisfied or puzzled by it? If the author started out to prove something, did he or she fulfill your expectation? If the author started out to analyze something, did you feel he or she lived up to the commitment? Did you come away feeling the author knew the subject well, presented it well?

Evaluating Style Style is how the author writes, how he or she uses language. Theoretically, style can't be separated from content (if you could subtract content, there would be nothing left), but for purposes of analysis, we can talk about word choices, sentences structures, and so forth. Style is a very subtle kind of evidence, easy for the writer to manipulate, hard for the uncritical reader to detect.

1. How readable is the style? Big words, long, complicated sentences, and abstract concepts aren't necessarily bad—unless there are too many of them. How easily can the educated general reader (you) read this language? What evidence is there that this writer has taken care to make the writing clear, simple, and easy to read (relatively)? No one expects adult researchers to write in simple-minded "Dick and Jane" sentences, but some research is a good deal easier to read than others. Some research

is so full of technical jargon and polysyllabic words that the research is nearly useless—unreadable except by a very few experts who are willing to work hard to decipher the writing.

2. What is the tone of the article? *Tone* is the author's attitude toward the subject: Is it dry, objective, bland? Would you say the tone is dull or boring? Has the author attempted to use tone to interest the reader? Is the tone lively, entertaining? Where are the clues to tone? Research is not often jolly nor deliberately written to entertain the reader. But researchers have an obligation—just as other writers do—to assist readers (versus anesthetizing them). Long, rambling, ponderous sentences; too much repetition of words and ideas; too much unnecessary digression—all have negative effects on research.

   Is the language appropriate? Can you distinguish between unnecessarily complex language and necessarily complex concepts? If the language is too hard or too simple, why is it? Who is the intended reader? (Scientists? Children?) Does the author treat dignified subjects with dignity and humorous subjects lightheartedly? Is the author sarcastic; does he/she show off with language; is he/she dull, unclear, or in any other way using inappropriate language?

3. Does the author have good diction (vocabulary)? Can you find specific words that look especially well chosen, accurate, colorful, effective? Does the author use figurative language (metaphor, simile, analogy, rhetorical devices) or is it all straight, literal reportage? Are there any surprises in vocabulary, or is it all pretty humdrum and uninspired—is it supposed to be? Are any new words used, words unfamiliar to you, to the general reader? You need to know exactly (not sort of) what is meant by *offal*, for example. When you aren't sure of the author's meaning, write a question to yourself: Why does he say this? What does he mean by *existentialism*? Whom is he calling "disloyal Americans"? Are new words good or bad here? (Too many may make the thing unreadable, but a few new ones might make it more interesting: What's your judgment on it?) If the author introduces technical terms, does he or she explain them?

4. Does the author write good sentences? Are there any especially memorable ones, well worded, capturing the ideas effectively? Any variety in sentences, surprising phrases, clever or helpful signals to the reader? Does it seem as if this writer took the time to polish and write skillfully, or does it all sound pretty bland and ordinary? Any bad sentences—confusing, awkward, fuzzy wording? Too many short sentences? Too many long ones? What effect does the quality of the writing have on the reader's reaction to the information?

5. Is the article structured effectively? Do the sentences add up to good paragraphs (that can be followed and understood by an educated reader)? Are there transitional devices to help the reader along? Has the writer

used subheadings, charts, diagrams, tables? Is the material organized in some logical fashion, or thrown together in random order? Is this organization calculated to look objective, free of bias, "natural"? Or would you say the structure is "loaded" to favor one side or the other? Can you describe the organization? That is, is the organization of the writing apparent to the reader?

6. Can you detect the author's attitudes, either about the reader or about him- or herself? Who does the author think he/she is? Very Important Person? Just Plain Jane? Can you relate to this author's "voice" (self-concept)? Who does the author think you are? An expert? John Q. Public? Is the author condescending to the reader or writing over your head? If you decide the author is offensive in some way, do you think the offense was deliberate? If you can't detect the author's attitudes, is that good? Does it mean the author is being objective? (Or does it mean the author is just bland, toneless and dull?)
7. What kinds of appeals to the reader does the author use? Ethos? Logic? Factual data? Emotion? Do you accept these appeals as legitimate, or do some of them offend you (see section on Fallacies in Chapter Four). Where are the clues to the author's appeals?
8. How does the author's style affect your reaction? Is the style one you admire, one you can accept as fair, reasonable, authoritative? Are there distortions of style? Does the style conflict with the message? Do you think there is no style? Where are the clues to the author's style?

   As a writer yourself, did you find the writing enjoyable? Were you able to learn anything from the author's techniques? How would you describe the author's prose—every sentence polished and pleasing, loaded with information? Is this an article you think other researchers should read? Does your reaction to the quality of writing have any influence on your judgment of the quality of the content?

Style and content operate together, never separately. The same words that express information also express attitudes, assumptions, qualifications. With a slight shift in emphasis or connotation, writers can alter the meaning of a sentence. By rearranging the organization of a text, a writer can influence the reader's interpretation. By understating, or not stating at all, a writer can bias the reader's understanding. The closest thing we have to "pure" information is a telephone book or a dictionary, and even there we can have differences in style (compare a child's dictionary with the adult version). Style has more to say about content than most people realize.

When we evaluate a text, we try to estimate its value—how good it is. Is the information good information? Important? Useful? Worth having? When we read a new research study, we want to know how good the study is, how reliable the information is. However, we can't judge content without also judging style. The reliability of a study depends a great deal on how the study was designed, conducted, and reported. Too much of our research in the past has become dusty books and journals forgotten in our libraries because of poor writing. Good writing can't magically compensate for faulty research, but poor writing will almost certainly corrupt good research. It's important then, especially for students, to pay critical attention to both content and style when evaluating research.

ACTIVITY 23 Analyze the content and style of the following article.

MARINE POLLUTION IS KILLING THE OCEANS AND LAKES

When thinking of the oceans, we envision clear blue waters and gently rolling waves. Unfortunately, this picture is rapidly changing. All over the world, people are abusing the oceans and lakes, and the waters of the earth respond by changing. These changes are just beginning to be noticed. Scientists are just starting to pursue them. The ocean does not yield neat bundles of statistics and absolute facts on the state of its health.

Nevertheless, the ocean has suffered immensely from these few years of man's intervention. People are dumping solid wastes into the waters, industrial sludge and the byproducts of domestice sewage treatment (Williams, 1979, p. 23). Heavy metals (pesticides, petroleum products) are also dumped into our waters (Rygg, 1986, pp. 33-34). Most deadly of all are the extremely toxic radioactive wastes from nuclear power plants, nuclear fuel production plants, and reprocessing plants (Williams, 1979, p. 26). Every imaginable filth and poison is freely dumped into our waters, entering the food chain. In 1985 the amount of garbage, dredge and oil were accumulating in such large amounts that they were actually beginning to form islands off the coast of Japan in Tokyo Bay (Thompson, 1985, p. 431).

People use the oceans as they always have, freely and confidently. But what it is being used for has drastically changed. Instead of being a place which provides food and a place to swim and bathe in, it is now becoming a cesspool. A festering dump where industries and people selfishly get rid of their garbage, failing to think of the consequences.

Although there has been a tremendous number of warnings in the last decades that these new uses are terribly damaging to the oceans, we still cling to the now false comfort of thinking that nothing we do will harm the ocean—even when the evidence is visible.

For example, in 1986, 130 packages containing toxic, inflammable, or other hazardous substances washed up on the beaches of the New England shores (White, 1976, p. 156). White stated that these packages caused eighty people to be hospitalized. Many other people were also affected by symptoms that were not severe enough to require medical attention. These ordinary people were simply taking advantage of a natural resource, the ocean waters, by swimming in them, something that had always been done, never realizing that it could be harmful. If pollution continues, everyone will be afraid to spend a leisurely family day at the beach.

Not only are our oceans being killed by pollution, but our fresh waters are also being affected. Bower (1985, p. 188) found that swimmers in the Lake Huron, Lake Ontario area have an illness rate of 40% more than nonswimmers. These illnesses include respiratory, gastrointestinal, eye, ear, and skin infections that are caused by staphlococcus and intestinal bacteria (Bower, 1985, p. 189). The source of these bacteria, as stated in Bower's 1985 study, is mainly sewage, although pathogens shed by bathers, low rainfall, and crowded beaches make the problem worse. It is sad to think that even our freshwater lakes are unsafe to swim in.

The worst polluter of all, and the one that unfortunately is dumped in the ocean the most, is oil (Simon, 1984, p. 54). Oil is also very evident and visible to the public, especially to coastal countries. In 1985 there was so much oil surrounding the coast of Japan that the water actually appeared to be brown (Thompson, 1985, p. 432).

Oil is a very dangerous substance to be dumping into the ocean. Unlike other poisonous pollutants such as DDT's and PCB's, which first of all attack the microscopic plankton, oil deforms the larger species, such as fish eggs, larvae, protozoa, and other marine organisms. Oil contains carcinogenic hydrocarbons, which are cancer causing substances that are incorporated into human food sources (Simon, 1984, p. 157).

The worst aspect of oil pollution is that it sinks into the ground sediments. Therefore it kills everything on the bottom of the ocean floor. Then it oozes out of the bottom, affecting everything above it. The sad ending is that the human population could not survive in its modern world without oil, which means it must be disposed of somewhere.

Industries choose the ocean to dump oil and other pollutants because it is probably the cheapest place, and they advertise to the world how safe it is, since it is dumped so far away from people (Barnett, 1986, p. 86). But the bottom line is that it is not safe; the ocean does not kill the pollutants, nor does the bacteria die at sea. Instead it becomes dormant and survives quite well. By the time it reaches the coast, the waters are nothing more than a sea of pathogens.

The oceans are not yet dead from pollution, but they are very sensitive to the effects of pollutants. Its systems are slowly degenerating, its species are disappearing. As long as we find it necessary to use the sea as a global toilet, it will continue to slowly deteriorate until it finally perishes completely. The world must be taught how harmful marine pollution really is before this happens. Soon it may be too late.

References

Barnett, R. (1986). Burning wastes at sea. Technology Review, 127, 85–87.

Bower, B. (1985). Jaws of a different color? Science News, 128, 187–188.

Rygg, B. (1986). Heavy metal pollution and lognormal distribution of individuals among species in benthic communities. Marine Pollution Bulletin, 71, 31–36.

Simon, A. (1984). Neptune's revenge: The ocean of tomorrow. New York: Franklin Watts.

Thompson, M. (1985). Islands in the waste stream. Science, 229, 431–432.

White, G. (1987). The dangers of flotsam and jetsam. Environment, 29, 21–22.

Williams, J. (1979). Introduction to marine pollution control. New York: John Wiley and Sons.

Barbara Zielinski

ACTIVE READING

Reading research is much like studying. It's a very purposeful kind of reading. Researchers don't usually have time to dawdle or linger over a text; there is so much to read that you must develop procedures for handling all the data as efficiently as possible. There are exceptions to this, of course. If you are reading poetry, fiction, or philosophy, one of your readings ought to be slow and perhaps lingering. But in general, you need to be able to skim and read as rapidly as you can, taking care to make sure you aren't missing anything important nor distorting what you are reading. It will do you little good to zip through books and articles at tremendous speeds if you can't understand what you're reading, can't find all the pertinent information, can't take thorough and accurate notes.

Surveying the Material

Before reading a book or research text, survey it. (Because "research" can include books, reports in journals, magazine and newspaper articles, government documents, and a host of other kinds of materials, I use the word *text* to mean any kind of printed research material.) Examine it. Determine as much as possible about it before reading.

- Think about the title: does it have a meaning? Why is the title worded that way?
- Look through the table of contents; fix in your mind what the book covers.
- Skim through and read preface and foreword material, skim the introductory and closing material. Pick up as much as possible about the book by skimming through it rapidly.
- Be sure to check for things like an index, pictures, charts, diagrams, tables, and the like.
- Try the index test: Think of three or four things you believe ought to be in the book, then see whether they are in the index.

The First Reading

The first reading should be rapid. Read noncritically, but pay attention to headings and labels, as well as to other guides to the content and structure. The first reading should be an unbiased, open-minded look at what the author has to say. That is, you owe the author a "fair" reading; the author must be allowed to make his or her point in the way that he or she wants to make it. Before imposing your analytical or critical judgment on a writer's work, you must read at least once, giving the writer the benefit of the doubt. Before making any judgment, listen to what the author has to say.

- After the first reading, write in your journal or notebook your overall impression of the book, what it's about, important points you recall.
- Summarize the book.
- For a thorough job, make an informal outline.
- This isn't the time to begin taking notes for research purposes. Instead, after the first reading (some researchers can do it during the first reading) your notes should be reminders to yourself, your own ideas, and reactions to the book.

Sometimes the first reading is all you need. If the material is especially simple or not particularly useful, you may not need to read further. Or you may have discovered on your first reading that there are only one or two things in the text you can use, and all you need is to go back and take notes on those specifics. More often the first reading merely prepares you for the second reading. You now know what the text is about, what it covers, how it's organized, and how difficult it is. With this information you can proceed to the second reading.

The Second Reading

The second reading should be relatively slower, much more deliberate. In general, always read as rapidly as you can; faster reading is often better reading, research shows, and there is less mental fatigue from faster than from slower reading. But almost always, relative to the first reading, the second is slower. Always speed up when you find the material permits that; slow down when you must. The second reading should be research reading, and the notes you take should be research notes.

Mark the Text

For the second reading, you may wish to use an underliner or highlighter—but sparingly. Mark the main points only; try not to mark things on the basis of interest. Some readers mark only what interests them personally, but that produces only a biased, personal interpretation of the text. Instead, mark the structure of the author's ideas (main points). As you read, change titles and subheadings into questions. Change "The Three Causes of the Civil War" to "What were the three causes of the Civil War?" You must think like a writer: What is this writer telling the reader?

Make Marginal Comments

Marginal comments are often better than highlighter markings. Marks aren't notes, and they don't call for much thought on your part. When you begin commenting on the text, responding to what the writer is saying, you are reading most actively. Of course, some marginal comments may still be in the form of "marks"—the question mark and exclamation point are popular marks, as are asterisks and arrows. These marks are reminders to yourself of how you responded at certain points in the text.

Question the Author

Don't read passively, accepting whatever the author says. Read actively. Ask questions, especially when you don't understand or you find the author unclear. In addition to or instead of marginal comments, you must write notes. A note should be a message from you to yourself. Dozens of underlinings and highlightings really aren't notes—they are more like signals saying "look here," but they give no clue to your thoughts. A note should be a thought you have about what you are reading. As a writer yourself, you can use what you know about writing to help you understand and respond to what you are reading.

The most important question is So what? You must make sure you understand implications. Sometimes writers only imply without really stating the conclusion they want the reader to draw. If the writer says, "The Reagan administration piled up the greatest debt in all of history," you must ask, "So what?" What does the writer want you to conclude?

Don't skip over difficult concepts, unfamiliar allusions, and figures of speech. Often the writer uses an allusion in an effort to help the reader. Figures of speech are meant to provide comparisons for the reader. Look them up. You must remember that the writer is trying to tell his/her readers something. It's plain enough that the author says, "The citizens of Agroville were very proud of their new grain silo." But is there something more here? Is the author suggesting that maybe they should not have been proud? Is the author being sarcastic or ironic? Why call it "their" silo when earlier the author states it is owned by the Grain and Feed Company?

If an author makes references or allusions to people, places, or things you're not familiar with, you must not skip over them. Reading is one of the best ways to educate yourself; look up biblical references, check out facts and figures, find out who Dr. Mengele was, what the Enola Gay was, where or what El Dorado was supposed to be. Educated readers and writers need a common base of knowledge; such shared knowledge allows shortcuts in explanations and provides depth and breadth to information. When an author makes a comparison—calling someone an "Ancient Mariner," for example—more is implied than just an "old sailor." Look it up. "The Ancient Mariner" is a poem with mystical associations. The author assumes the reader will catch not only the literal meaning of the term but the associations that go with it.

Read for meaning. It's never enough to be able to say what the writer said. The question is, What did the writer mean? Is there a meaning? Almost always there is, even in nonfiction writing. Paraphrase—one way to determine the writer's meaning is to attempt to paraphrase his/her ideas. (See the section on Paraphrase later in this chapter.)

Be sure to note also what the author does not say. Authors sometimes don't look at both sides of an issue. They sometimes don't reveal their own biases and assumptions. They sometimes fail to mention contradictory evidence, or they omit facts and arguments unfriendly to their thesis. Obviously, the more you know about the subject before you read anything, the more critically you can read.

Argue with the Author

Most people read passively, accepting anything an author says. Or, if the author says something outrageous, they simply dismiss it. Some people read so uncritically that they are unaware when an author is being inaccurate, using faulty logic, or even deliberately trying to mislead the reader. But you should read the way you wish to be read—actively, not only listening to the author but responding. Treat the written word as the author's half of a dialogue; you supply the other half. Don't just passively drink in the author's ideas; argue with the author.

When an author says something you doubt or find hard to believe, ask yourself, "How can this be true?" If it's a factual matter, you can try to check up on the author. If it isn't factual but conceptual (the author has proposed a puzzle or contradiction, perhaps), you can try to deduce the author's reasoning:

> If this is true, as the author suggests, then what follows? Possibly *A* follows. But if *A* is true . . . then what follows? Possibly *B* follows . . . and so on. But suppose not *A* but *Z* is true . . . then what follows?

Keep on this way until you are satisfied that you understand or that you have at least come up with a plausible explanation of what the author meant. Occasionally you may decide that none of your possible suggestions make sense, that nothing sensible follows from any of your interpretations. Then you have either discovered a flaw in the author's reasoning or have left yourself with a question that needs research.

Use your imagination; permit yourself to daydream a little, speculate, digress from the text. Suppose the author is right—what then? Perhaps the author proposes a future that will be free of disease—think about that for a minute. What sort of world would that be?

ACTIVITY 24 Read the following article. Use your reading journal to react to the article: Read actively. Follow the suggestions in this chapter for first and second readings.

IS NIM, THE CHIMPANZEE, PROBLEM SOLVING?

Summary.—It is argued that the importance of Project Nim has been exaggerated. The project contains many flaws such that the performance of the chimpanzee, Nim, provides very little information on a chimpanzee's ability with language. Nim learned exactly what he was trained; he was provided little opportunity either to learn or use language.

Project Nim has become one of the more influential of the chimpanzee language projects. The impact has largely been negative in that less of this type of research is currently being undertaken. The results of the project have been used in criticism of other attempts to train chimpanzees' language. The author argues that the project is flawed and its importance has been exaggerated.

The results of Project Nim were that Nim learned Ameslan at the same approximate rate as previous chimpanzees; in 46 months of training Nim acquired the productive use of 125 signs. Nim also produced more than 19,000 different multi-sign sequences in the first 40 months of the project. Nim's two-sign combinations appeared to show the same semantic relations as those identified in other projects.

In many ways the results appeared to support the contention that chimpanzees are capable of language-like performance. The most compelling reasons for a negative conclusion were provided, however, by a discourse analysis of Nim's signing (Sanders, 1980). Sanders discovered from this analysis that many of Nim's signs were simple imitations, that Nim's non-imitative signs were independently produced labels of the environment and that Nim had a series of ''wild cards'' that could be used in any situation; for example, ''me,'' ''Nim,'' and ''more.'' Terrace, Petitto, Sanders and Bever (1979), therefore, suggested a non-relational model of Nim's language. They suggested that Nim's choice of signed combinations was not determined by any semantic relationship among elements of the combination. The importance of this result, however, has been overestimated.

There are at least three possible reasons for a negative result in research on language by chimpanzees: a failure in the training procedure, a failure in experimental design (Muncer & Ettlinger, 1981), a genuine inability on the part of the chimpanzee. It appears that Nim's failure may have been occasioned by the first and second sources.

The number of trainers used in the study was extremely large and may have limited any linguistic performance, as Fouts noted:

> Communication is the binder of a relationship. By having so many changes, Terrace de-emphasized the relationship necessary for language to develop (Fouts, see Marx, 1980, p. 1331).

Terrace also commented:

> His emotional reactions to the steady replacement of volunteer teachers suggests that his use of sign language may have been limited as much by motivational as intellectual factors (Terrace, 1976, p. 66).

Fouts argued that by relying on operant conditioning Terrace would inevitably produce a passive animal whose behavior is largely imitative (Marx, 1980), but this does not necessarily follow from the use of operant conditioning. Nim's failures reflect the poor use of operant conditioning rather than its use *per se*. The training methods appear inconsistent and inadequate. Sanders (1980) had admitted that he could not be sure which methods were used and with what frequency. The reinforcement contingencies also remain a mystery:

> However, teachers probably used their judgment in deciding when to make

> reinforcement available on a continuous schedule and when on an intermittent schedule (Sanders, 1980, p. 23).

From the sparse information given, it is apparent that Nim was encouraged to believe that a wide variety of arm movements would be appropriate in the same situation. For example, as a request Nim could sign ''Nim,'' ''me,'' ''eat,'' or the name of an object. This training would encourage a non-relational approach to language and also encourage the use of the demand mode. As much is concluded by Sanders when he said that:

> This signing strategy (the non-relational model) is similar to the way in which Nim was encouraged to use signs during vocabulary training (Sanders, 1980, p. 99).

The approach to syntax may also have been unwise. The trainers were told not to respond differently to combinations nor to try to train combinations. It is known that chimpanzees have to be taught language specifically, unlike children. It also seems likely that syntax evolves naturally in children as a result of the emphasis on communication (Newport, 1977). It is not likely, therefore, that Nim would naturally be able to abstract regularities in language without either encouragement to do so or an emphasis on communication. Attempting to teach Nim to use combinations would not have affected the relevance of his performance, it would still have been possible to determine whether Nim had merely understood lexical ordering. It is not surprising that Nim failed to learn to use order in any interesting way, given the ambiguities presented in training and the lack of need for him to do so. Terrace (1981) recognized as much when he said:

> Why should an ape be interested in learning rules about relationships between signs when it can express all it cares to express through individual signs? (Terrace, 1981, p. 112).

Project Nim is in many ways commendable, particularly because discourse was analyzed. A brilliant analysis of flawed data, however, does not help to clarify the issues or give answers. Project Nim is merely another demonstration of the difficulties in training a chimpanzee.

As far as apes' language is concerned the data are definitely not in, and a few negative results and errors solve very little. At the present time there is some evidence that chimpanzees are capable of learning syntactic rules when trained (Muncer & Ettlinger, 1981; Passingham, in press). One chimpanzee, for example, has shown ability to change the order of prepositional sentences in response to changes in positions of objects and also ability to understand different meanings when prepositional sentences were presented in different orders. There is also evidence of a chimpanzee's ability to provide information to another chimpanzee and to classify objects according to semantic category (Savage-Rumbaugh, 1982). It seems likely that the ability of a chimpanzee to learn language may have been overestimated by some, but their abilities can also be underestimated by using an imperfectly designed training program.

References

Marx, J. L. Ape language controversy flares up. Science, 1980, 207, 1330–33.

Muncer, S. J., & Ettlinger, G. Communication and syntax in the chimpanzee: First-trial mastery of word order that is critical for meaning, but failure to negate conjunctions. Neuropsychologia, 1981, 19, 73–78.

Newport, E. L. Motherese, the speech of mothers to their young children. In N. J. Castellon, D. B. Pisoni, and G. R. Potts (Eds.), Cognitive theory. Vol. 2. Hillsdale, NJ: Erlbaum, 1977. 46–72.

Passingham, R. E. The human ape. London: Freeman, in press.

Sanders, R. J. The influence of verbal and nonverbal context on the sign language conversations of a chimpanzee. Unpublished Ph.D. dissertation, Columbia University, 1980.

Savage-Rumbaugh, E. S. Acquisition of functional symbol use in apes and children. Paper presented at Harry Frank Guggenheim Conference on Animal Cognition, Columbia University, 1980.

Terrace, H. S. A report to the academy, 1980. Annals of New York Academy of Sciences, 1981, 364, 94–113.

Terrace, H. S., Petitto, L. A., & Bever, T. G. Project Nim: a progress report 1 and 2, Columbia University Psychology Department, 1976.

Terrace, H. S., Petitto, L. A., Sanders, R. J., & Bever, T. G. Can an ape create a sentence? Science, 1979, 206, 891–902.

TAKING NOTES

Some students have the habit of photocopying entire articles or chapters in books without reading the material first. This practice merely delays the actual work of note-taking and is an ineffective way to go about research. Occasionally you may find an important article that you would like to copy, but most library material can be handled in a simpler fashion.

It's important to create a system of note-taking; relying on random scraps and bits of information makes your job too hard. Serious researchers devise a uniform method for note-taking as part of their procedure for data collection. The most popular method is the use of note cards.

You need two kinds of cards: one for sources (bibliography cards) and one for data (note cards). The bibliography cards are your control cards. These can be shuffled about, arranged alphabetically, and supplemented easily. One good reason for keeping the bibliography on cards is so that you can continue to add to them. At the last minute you may find an important article. However, if you are using a computer (instead of a typewriter), you may find it's just as easy to keep your bibliography in a file, which you can update and print out whenever you need it.

On a 3-by-5-inch card write out the full bibliographic entry exactly as it will appear in your paper. See Figure 2.6. Always give the call number. Place a control number in the right-hand corner. You may arbitrarily assign control numbers in the order you find your sources: the first one becomes number 1, the second one becomes 2, and so on, regardless of alphabetical order. The control number will allow you to match note cards with bibliography cards.

As you begin to read your materials, you must keep in mind that you are a researcher, not a recorder. The object is for you to understand the information, not just to transmit it from the library to your research paper without passing it through your mind. In the library, read much, write little. Copying

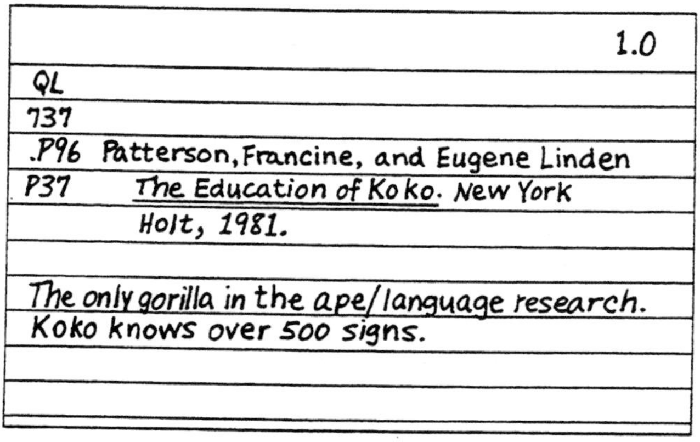

1.0

QL
737
.P96 Patterson, Francine, and Eugene Linden
P37 The Education of Koko. New York
Holt, 1981.

The only gorilla in the ape/language research.
Koko knows over 500 signs.

FIGURE 3.1 A bibliography card (3″ × 5″)

volumes of notes, even with the aid of the nearest copying machine, is a waste of time. You must still *read*, *understand*, and *extract* the information you need. Copying without understanding can lead to problems, as when later paragraphs contradict the one paragraph you copied because it seemed to support your thesis.

If you do copy, copy exactly. At the outset it's important to keep clear which are your words and which are in the source. Put quotation marks around copied material just as you would in your paper. If necessary, add explanatory remarks that will help you identify the original (for example, "Quotation marks in original"). If you paraphrase or alter the source material in any way, leave yourself a comment explaining the change, for example, "My paraphrase." Some researchers believe it's so important to keep straight who said what that they put comments like these in the research report itself.

Most of your notes should be messages to yourself. You can note whatever you think is important about the source without copying it. There is a research bias against too much use of quotation. Stitching quotations together isn't research. (See "The String of Pearls" in Chapter 5.) You are expected to assimilate the material; thus, most of your use of source material should be in the form of paraphrase, not quotation. Remember that you are supposed to read and *understand* the source. Think about what the data mean and where they fit into your research, and then jot down only brief notes. You may need larger cards for notes than for bibliography, but the reason for using cards instead of sheets of paper is to limit the amount of writing. For the same reason, it's not a good idea to use your computer for taking or keeping notes. The computer encourages copious writing; its theoretically limitless pages (disks) allow you to ramble on and on, copying and paraphrasing without thinking. During the note-taking cycle, the less you write the better; read much, *think* much, but write little. Each card should contain only one fact, idea, quote, comment, or remark. The 5-by-8-inch card is the most popular size. Figure 3.2 shows an example of a note card.

Model Notes

It isn't possible to describe all the various types of notes you might put on your cards. Most researchers eventually develop their own methods. In general, there are the following possibilities:

DESCRIPTION OF SOURCE

Wolfe's The Right Stuff is an example of journalistic non-fiction. It is the history of America's space program from its beginnings about 1955 to approx. 1965. It particularly concerns the first astronauts. The title refers to courage and daring: being made of ''the right stuff.''

This kind of note is useful for a source whose relevance is uncertain. It's useful for a source that merely confirms or adds to a point already well established with other sources. It's also useful for a source whose relevance is simply that it exists—in this case, an example of modern journalistic nonfiction. (That is, if you wanted to name a few journalistic works, one you could name is *The Right Stuff*.)

EVALUATION OF SOURCE

Blain's news items, about ''monkeying around with the language'' is a short, undocumented humor piece in which she treats the idea of apes using language as ridiculous. Very light article, no substantive data here, but might be useful as illustration of public attitude toward issue—i.e., nonserious.

Evaluating a source is always a good idea, particularly when the source is of questionable value. See Rules of Evidence, Chapter Four.

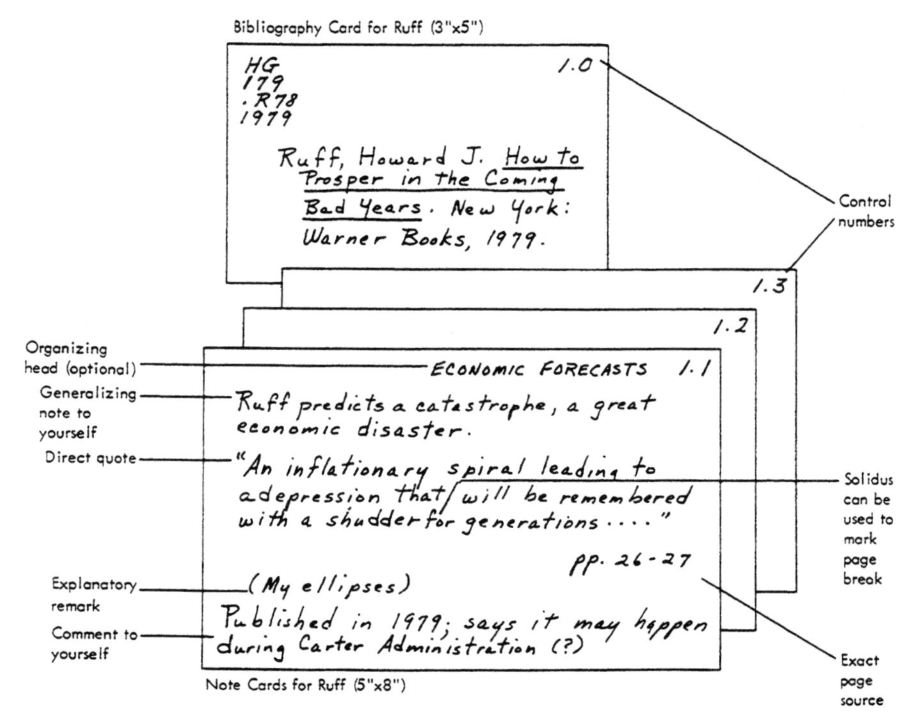

FIGURE 3.2 Note cards (5″ × 8″) and a bibliography card (3″ × 5″)

SUMMARIZING

There are different kinds of condensations called summaries. Some researchers insist on maintaining the distinctions among the types. Some teachers insist that the differences are absolute. For your private purposes as a researcher, you can decide which to use on the basis of whichever seems most appropriate. (See Appendix.)

The Summary The summary is an informal condensation, in your own words, of source material. In it you try to capture ideas and reduce the source material to its bare message. The summary may be a single sentence or several sentences, as you think appropriate. For example:

> Romeo and Juliet are young lovers whose feuding families would not approve of their relationship. In a series of mistakes, Romeo and Juliet both commit suicide. All ends in tragedy for the feuding families.

The Précis and the Abstract A précis (PRAY-see) is a formal, careful summary that preserves all key points and all significant subpoints. It's virtually a sentence outline in paragraph form. The older word *précis* is now frequently replaced with *abstract* in scientific writing. Both mean a careful summary of greater or lesser detail, depending on your purposes, who your reader is, what your writing situation is, and so forth.
Example:

> Romeo and Juliet fall in love before they learn each other's names. His is Montague, and hers is Capulet; they are teenagers of warring families. In a street fight, Juliet's cousin Tybalt kills Romeo's friend Mercutio. Romeo then kills Tybalt and is exiled by the Prince (of Verona), Mercutio's kinsman. Juliet's father arranges that she shall marry Paris, a young nobleman, kin to the Prince. To prevent this marriage, Juliet arranges with Friar Laurence to take a potion that will seem to kill her. A message is sent to Romeo to meet and elope with Juliet after her ''funeral.'' The Friar's messenger misses Romeo. Instead, Romeo's own servant reports the ''death'' of Juliet. Romeo rushes back to Verona, finds Paris at Juliet's crypt, and kills him. Romeo finds Juliet's ''dead'' body, and in despair kills himself. Juliet awakens, discovers Romeo's death, and kills herself. The Capulets and Montagues are united in grief and pledge peace between their families.

MAIN POINT OF BOOK, ARTICLE, PASSAGE

The main point of Ferrin's article is that the price of gold is now falling because investors can make more money by investing in high-interest issues. He says gold will continue to fall but that it will rise again, eventually hitting $1,500 per ounce!

RELEVANT STATISTICS

In addition to the 100 senators and 435 representatives, there are about 18,000 employees of the congressmen: approx. 27 for each representative and 65 for each senator. There are also some 15,000 registered lobbyists—with their support personnel. (p. 279)

This information, from *Free to Choose* by Milton and Rose Friedman, contains statistics that would have to be documented (attributed to their source) if you were to use them. Thus the page number is required. With it and the bibliography card for the Friedman book, a researcher has at hand all the information that goes into a parenthetic citation and a Works Cited page.

GOOD EXAMPLE OR ILLUSTRATION

The following example includes a page number, as does the preceding one, because a note-taker would have to document the direct quote if he or she were to use it in a paper. (The quotation is from *The New Journalism* by Thomas Wolfe and E. W. Johnson.)

Wolfe's narrator uses the downstage voice, creating the illusion of the narrator as participant (instead of reporter, after the fact):

> ''Working mash wouldn't wait for a man. It started coming to a head when it got ready to and a man had to be there to take it off, out there in the woods, in the brush, in the brambles, in the muck, in the snow. Wouldn't it have been something if you could just have set it all up inside a good old shed with a corrugated metal roof and order those parts like you want and not have to smuggle all that copper and all that sugar and all that everything out there in the woods and be a coppersmith and a plumber and a cooper and a carpenter and a pack horse and every other gaddamned thing God ever saw in the world, all at once.'' (p. 18)

PARTICULARLY WELL-WORDED, MEMORABLE STATEMENT

Extremism in the defense of liberty is no vice. And . . . moderation in the pursuit of justice is no virtue.

[Barry Goldwater, Acceptance Speech, Republican Presidential Nomination, July 16, 1964.]

SIGNIFICANT ERROR, CONTRADICTION, PUZZLE

Although he's been dead for 19 years, thanks to the dedication and determination of one Rochester woman, Edgar Guest has not gone unforgotten. (?)

[The Rochester (MI) Clarion, 8/7/80.]

INFORMAL OUTLINE

[Major Points]
''The navy has discovered a rationale for unlimited expansion: the protection of imported raw materials.''

[Subpoints, with or without numbers]

1. Persian Gulf crisis showed need for protection of oil supplies.
2. Minerals like titanium, cobalt, and so forth are scattered across the globe and could be crucial to our survival in the future.
3. Role of navy in protecting imports means increased buildup of ships.

But Klare says Pentagon claims are exaggerated.

1. U.S. has large reserves of most minerals.
2. Many critical minerals can be obtained from Canada, Mexico, Australia—''friendly'' countries.

There are alternatives to military protection.

1. Additional stockpiling.
2. Development of substitute materials.
3. Development of alternative sources (most raw materials can be found in more than one place in the world).

Conclusion: There is almost no way to protect raw materials from terrorists, war, unstable governments, and so forth. Reducing our dependency on imports is a better solution than spending billions trying to prepare for a ''resources war.''

[Michael T. Klare, "Resource Wars," *Harper's*, Jan. 1981: 20–23.]

The informal outline is written for your private use and can therefore be quite informal. Whether you use numbers and letters is up to you, as is how much detail you put into the outline. You can see that most of the clarifying and illustrative details have been omitted from the example here. How much detail you go into in an outline depends on your purpose in outlining. If you find an article extremely useful, you may want to prepare a highly detailed outline. An outline is one of the most effective aids to analysis and understanding you can use. In the early stages of your research, you may find that outlining articles gives you the greatest insight into your subject.

The outline reveals the *structure* of your study material. It shows not only what the important ideas are, but also how they are related to each other. Writers construct their outlines while developing their compositions; ideally you, the reader, can reproduce that outline when studying. The outline reveals that the material is not a list, not a random collection of ideas—it's an organized whole.

PARAPHRASE

To paraphrase means to state an author's meaning in your own words. Paraphrasing is an excellent way to discover the meaning of a passage. To paraphrase, you must think about what the original words mean and try to find your own words to state that meaning. As you sift ideas and reject some words in favor of others, you will be actively engaged in thinking about your reading.

> ORIGINAL
> There is difficulty in praising convincingly an achievement with which one has been associated. Architects are not thought best for reviewing their own buildings, poets their own verse.
>
> John Kenneth Galbraith, *A Life in Our Times*

> A PARAPHRASE
> If you praise some accomplishment you had a part in, people may assume you are flattering yourself.

Note that this is only *a* paraphrase and not *the* paraphrase. Because we are dealing with the meaning of Galbraith's statement, there is room for greater or lesser degrees of accuracy.

To paraphrase does not mean to "translate" with synonyms. In an early effort to create a computer that could translate from English to Chinese, the computer's translation of a familiar expression—"Out of sight, out of mind"—produced the Chinese equivalent of "absent idiot." The computer had literally "translated" the words, but it disregarded the meanings of the words in context. Paraphrasing doesn't mean saying what something *says* in different words; it means saying what it *means*.

Paraphrasing is an especially good technique when you are taking notes on something difficult; the act of paraphrasing will help you discover at least what you think the author means. And if paraphrasing helps you to understand, it will also help your readers understand. When you write something readers may have difficulty with, you can offer your own paraphrase: "What I mean by that is . . ." or "Another way to say that is . . ." and other expressions may introduce paraphrases.

ORIGINAL
We have nothing to fear but fear itself.

Franklin D. Roosevelt, 2nd Inaugural Address

BAD PARAPHRASE, LITERAL TRANSLATION
We are without anything to be afraid of other than fearfulness itself.

This isn't really a paraphrase at all; it's simply a substitution of synonyms for the words in the orignal. A paraphrase should be an answer to the question, "What does it mean?" Merely changing the words gets us no closer to an answer than the original did. Furthermore, the closer your paraphrase is to the original words, the more it will seem like plagiarism. Better paraphrases of Roosevelt's statement might be one of these:

Panic and destructive behavior resulting from fear are greater dangers than most of the things that cause fear in the first place.

The things people do when they are afraid are worse than the things they are afraid of.

The results of fear are often more destructive than the causes of fear.

Or, to state it in more positive terms: "If people remain calm, we will be able to survive the present danger."

ACTIVITY 25 Write a paraphrase of the following passage:

No man is an island, entire of itself; every man is a piece of the continent, a part of the main; if a clod be washed away by the sea, Europe is the less, as well as if a promontory were, as well as if a manor of thy friends or of thine own were; any man's death diminishes me, because I am involved in mankind; and therefore never send to know for whom the bell tolls; it tolls for thee.

John Donne, *Devotions*

STUDY GUIDE:
Chapter Three Reading Research

1. In what sense is reading self-education?
2. What are the limits on what a researcher must read?
3. What is meant by reading below the surface?
4. What is the role of purpose in reading?
5. True or false: Researchers should read everything slowly and carefully. Why?
6. What are four general reading purposes?
7. What does it mean to assimilate information?
8. What is meant by reading for analysis?
9. What is the simplest method of analyzing a research article or book?
10. What is parallel language?
11. How can researchers avoid single-item divisions in an outline?
12. What is rhetorical analysis?
13. What is the meaning of "voice" in research writing?
14. What is the proper voice for most research writing?
15. What is the relation between voice and credibility?
16. What is *ethos*?
17. What is *logos*?
18. What influence does the reader exert on research writing?
19. What is *pathos*?
20. What is the function of interpretive reading?
21. What is the purpose of reading to evaluate?
22. What is the "content" of a composition?
23. What is style?
24. What is tone?
25. What is diction?
26. What is active reading?
27. How can a researcher survey a text?
28. What is the function of a first reading of a text?
29. What is the purpose of marking a text?
30. What are marginal reading notes?
31. What is the most important question to ask about your reading?
32. What is an allusion?
33. How does paraphrase relate to meaning?
34. How can a reader argue with an author?

35. Why not photocopy material instead of reading it in the library?
36. What is the most popular method of note-taking?
37. What is the general rule on how much to write in your notes?
38. What problem may arise from copying without understanding?
39. Which is better while collecting data—copying or paraphrasing?
40. What else might appear on your note card besides copied material?
41. True or false: Most of your notes should be material copied from sources.
42. What is the difference between a summary and a précis?
43. What is an abstract?

CHAPTER FOUR

Evaluating Evidence

The key to all research is the evaluation of evidence. Collecting the data, taking notes, making outlines, using proper documentation, and writing in an academic style are all important components of research, of course. However, nothing is as important as the analysis and interpretation of the evidence. Suppose you find several authorities who say that copper bracelets help arthritis. What does this mean? Is this good evidence? Is the case proved? Suppose you find an equal number who say the opposite. Do the second authorities cancel out the first?

What is evidence? How do you evaluate it? The researcher is expected to analyze the data, weigh the evidence—not just compile it. The job of the researcher is to tell us what the data mean, to show us what conclusions should be reached from the data. It's the analysis of the evidence, not just the presentation of it, that tells us whether your research is significant.

PRIMARY AND SECONDARY EVIDENCE

Evidence is of two kinds: primary and secondary. Primary evidence is firsthand data, sometimes called hands-on data. It's data you find by your own efforts through experimentation, laboratory investigations, or fieldwork. Most beginning researchers don't have the resources for experimental research, but there are some possibilities you could explore. College campuses provide ready-made populations for surveys and questionnaires. For the same reason, you have the resources for simple experiments in human behavior, testing reactions to jokes, inkblots, and so forth. You may be able to think of a number of relatively simple but useful experiments to undertake—product comparison tests, for example, like those in *Consumer Reports*. You can test various brands of common products like soap, toothpaste, and chewing gum. A popular research project is the analysis and interpretation of a work of art such as a novel, play, or poem, in which you rely primarily on your own reading and understanding of the work to illustrate a thesis. Such projects constitute primary research.

However, almost all research makes some use of secondary evidence. You need secondary research even when conducting a first-hand experiment. Before you can start any research of your own, you must go to the library and find out what has already been done in the field you plan to study.

The library is well suited to secondary research. You can ask the basic question, What do researchers already know about the subject? Who said what? The standard procedure is to collect as much data as you can find on some question like "Should we register handguns in America?" and then to quote from representative authorities to show the distribution of evidence (some say yes, some say no). In such research you usually can't find one definitive answer to your question; instead, you will find various answers. This leaves the conclusion (*your* answer) up to you, as it should be in worthwhile research. You must *weigh* the evidence, *analyze* the data, *interpret* the findings. Then you must *decide* what your answer is. It isn't surprising, therefore, that many researchers decide the evidence is inconclusive. Many researchers conclude that there isn't enough data or that the evidence isn't good enough to give a clear, final conclusion. Such research is valid. It's better to know that we need more research than to mistakenly assume we have solved the problem.

CONVINCING WITH EVIDENCE

Research is partly science, partly art. It's not at all mechanical: the analysis and interpretation of data are intellectual tasks, both demanding and rewarding. Furthermore, writing readable, effective research papers amounts to a prose art. The skills for answering primary research questions are the skills of the laboratory. The skills of secondary research are the skills of the courtroom, and there is no sure-fire formula for determining the answers to courtroom or secondary research questions. Usually the best you can do is to give reasonable answers based on the currently available knowledge, knowing that later researchers may find reasons for discarding your answers. In general, in dealing with evidence, the following requirements are standard.

Consider the Weight of Evidence

The more evidence there is for one side of a question, the more believable that side will be. If ten psychiatrists find the defendant sane and two find him insane, the jury will believe the ten. However, the weight of research isn't the only consideration. All the researchers could in fact be wrong, or all partly right, or, as has happened in the past, the majority opinion could be the wrong one. One hundred experts swearing that apes can't talk may be refuted by a single talking chimpanzee. A single definitive piece of evidence can offset all other data. (That the accused was in prison at the time of the crime is a definitive fact that will offset fingerprints, motives, and a dozen eyewitnesses alleging that they had seen the accused at the scene of the crime.) Thus the question isn't automatically solved by discovering the distribution of the evidence. Nevertheless, it's important to discover what the experts think. Do most say apes can talk? Cannot talk? Is the break about even, half saying one thing, half saying the other? Is all the evidence of the same weight, the same quality? Or are some of the facts more convincing than others?

Consider the Source of Evidence

If two film critics disagree about a film—one loves it, one pans it—which should you believe? When a researcher is citing authoritative evidence from expert witnesses, it's important to establish credibility. If researchers cite published sources, how reliable is their evidence? What is the source of the data? Where does the evidence come from? Considering the source may or may not include the question of whom the evidence comes from.

Where Does the Evidence Come From?

If you're using oral evidence (interviews you conduct, for example), everything will depend upon your interviewing instrument and the sampling procedure you use. If you're using a survey instrument or a questionnaire sent through the mail, you must be aware of the problems in this kind of research. (Often fewer than half the sample group will respond; people fail to understand the questions; they tend to say whatever they think the interviewer wants to hear.)

It's difficult to say with certainty what constitutes reliable sources as opposed to unreliable ones. Researchers who use many "popular" sources such as *Reader's Digest, TV Guide*, the *National Enquirer*, small local newspapers, and reference works designed for mass markets should be prepared to defend the credibility of their sources (no paperback dictionary, for example, is as authoritative as *Webster's Third International* or the *Oxford English Dictionary*). Avoid digests and other sources in which information has been summarized or condensed. Always make sure you're reading the original source of information.

You may begin your reading with general reference works like encyclopedias; they provide summaries of large areas of research, and constitute what is known, in general, about any area. Your data collection should begin where the encyclopedias end. There is no point in "discovering" what is already in the encyclopedias; the information there can be considered common knowledge, basic information available to anyone. You, the researcher, are expected to offer something new. Therefore, it's a very good idea to get the overview, the background material in the encyclopedias before you begin your research.

There is some bias against quoting the encyclopedia or listing it as a source. Many researchers assume that you have read the encyclopedia coverage of your subject, but the information is too general and probably outdated by more recent research. It isn't safe to assume that encyclopedias are absolutely reliable and completely accurate. However, some of the better encyclopedias have signed articles by well-known authorities, and many encyclopedia articles contain bibliographies and source material you can make use of, especially in the early stages of your research. If you list an encyclopedia as a source (avoid those designed for children), be sure you have other sources to back up what you say. It's a good idea to have a variety of sources in any case. Unless the source bias is built in (if the subject is so new or so old, for example, that it isn't widely available), you should attempt to get a representative sampling of the kinds of sources available: books, magazines, journals, newspapers, government publications, and other forms.

Note that a too-heavy bias in favor of prestigious sources can lead you into the *genetic fallacy* (discussed later in this chapter), an assumption that the source of evidence automatically establishes its credibility. Research and discoveries at great institutions like Harvard and MIT, for example, aren't automatically more reliable than those from less eminent institutions. Large newspapers like the *New York Times* or the *Washington Post* aren't automatically more trustworthy than smaller papers. Research is quite impartial.

Who Gives the Evidence?

You need to evaluate the credentials of the authorities. Since most authorities have good credentials, students may be uncritical ("Who am I to challenge the experts?"). Furthermore—and unfairly, it would seem—students are required to have more documentation than the experts have. We are likely to accept the word of experts with less evidence than we require of students. In most serious research, however, the experts on both sides of any question are usually reliable. In that case there is little to be gained from splitting hairs over who has the most degrees, the most publications, the most authority. The credentials of the experts are important data but no shortcut for finding the answer to your question.

Nevertheless, you should be alert to some possibilities for bias in credentials. Powerful interests can exert pressure on research. It's naive to assume that objective research is free of the taint of money. Much so-called "objective" research is suspect because of the relationship of the researcher to those whose interests are either helped or hurt by the research. (See "Ad Baculum," page 131.) It's important to know who your authorities are, but famous scholars aren't automatically and invariably better than less-well-known researchers. A college freshman at a small school can, with careful and thorough work, produce significant research.

Authorities you cite in your paper offer different sorts of testimony. It's important that you know what sorts of testimony you are citing.

Statements of Fact One authority you are reading states that Lincoln lived for several hours after being shot by the assassin, Boothe. How does the researcher know this? Where did this piece of information come from? If the researcher is quoting someone else or drawing a conclusion based on evidence given elsewhere, this statement is hearsay evidence, and you should find the original source if possible.

Statements of Inference, Conclusion Inferences are often not verifiable from any one fact; they are conclusions based on the evidence. Examine the inference; find the evidence on which it is based. Do you agree that given the evidence the inference is plausible? (Washington was not revered in his lifetime?)

"Mere" Opinions "Mere" opinions are unrelated to particular evidence. The researcher is editorializing, offering his/her own ideas, not quoting secondary evidence. For example, "Coleridge's poem 'Hebe' is amateurish." You must ask, Who says so? If the writer offers substantiation, criteria by which you or I could agree or disagree with the opinion, it's no longer a "mere" opinion but a critical judgment.

Expert Testimony An expert is a witness with unquestioned authority. His or her knowledge is presumed to be technically accurate, that is, expert. Technical authorities are experts testifying about matters *on which* they are expert. Testimony from an expert witness is usually not considered "mere" opinion but a technical judgment; it has the weight of fact. Expert witnesses can be wrong, of course; occasionally one expert contradicts another, leading to the problem of whose testimony is the most expert. It's important for you to become as expert as you can when dealing with testimony from authorities. You need to know the subject matter well enough to understand the experts, and well enough to evaluate their testimony.

Consider the Evidence Itself

Assuming the authorities are all more or less reliable, your only recourse is to examine the evidence they offer. What kind of evidence is being offered? What conclusions are based upon it? If the prosecutor's psychiatrists have made only a brief examination of the defendant, the defense may successfully argue that its own psychiatrists have more persuasive evidence. If the prosecutor can then show that the defense's evidence was gathered under suspicious circumstances (the defendant's family has donated a large grant to the investigating institution, for example), the jury will be presented with the kind of argument it can handle—namely, disputes about evidence.

Prima Facie

(Prima FA-she) The general meaning of *prima facie* is "on the face of it; at first appearance." There is a *prima facie* case that parents haven't done a good job teaching teenagers about birth control: teen pregnancies. The meaning in law is "legally sufficient to establish a case." In both cases, *prima facie* tends to mean "obvious" or "undisputed." A body riddled with bullet holes is *prima facie* evidence of murder. The presumption may be wrong, of course; *prima facie* evidence argues only that on the face of it, murder seems a reasonable conclusion.

De Facto

De facto means "actually; in fact; in reality"; it's often used to contrast with appearances: "A *de facto* ceasefire existed even though both sides were still fighting vigorously." "The Arab army suddenly found itself the *de facto* authority over large parts of the Turkish Empire." In law the term *de facto* is used to mean "by order of law," "established or recognized by law": "In event of the president's inability to perform, the vice-president becomes *de facto* president."

Ex Post Facto

Ex post facto means "after the fact; retroactively." Anyone who assists a criminal fleeing the scene of a crime may be guilty of the crime *ex post facto*. Researchers in certain notorious cases were found to have altered their data *ex post facto* to agree with predetermined conclusions.

A Priori

The term *a priori* is commonly used to describe a mistake in which research (or argumentation) is undertaken to prove a conclusion; the researcher has decided in advance, for example, that the people of a certain jungle are "primitive and quaint" and undertakes research to prove that conclusion. Often these presumptions are entirely unspoken and unknown to the researcher (they amount to biases or prejudices). More formally it means "presumptively," arguing from definitions or assumptions. It was once possible to make an *a priori* argument that the earth was the center of the universe based on the observation of the senses, in which the sun and stars appeared to rotate about the earth. It tends to mean "without examination or analysis; independently of experience; intuitively, self-evident and therefore without appeal to the particular facts of experience; presupposed by experience." In philosophy, it's necessary to reason *a priori* in some cases, but on the whole *a priori* arguments are serious errors in research.

Ipso Facto

Ipso facto means "by the fact itself; as the result of the mere fact; by the very nature of the case." The term contrasts with *ipso jure*—"by law; by the operation of law." The fact that she has had a baby is *ipso facto* evidence that she was pregnant.

RULES OF EVIDENCE

Definition

The object of investigation must be clear. Are all the authorities talking about the same thing? If they are all using the same terms, do they all mean the same thing when they use them? A simple question like "Should heroin be legal?" can be loaded with many complications and subtleties: How much heroin? For whom? Under what conditions? Your thesis question may not be completely clear when you start your investigation, but it must become completely clear as you work on it.

Timeliness

How old are the data? Generally speaking, we believe that newer data are better than old, but this isn't always the case. Recent data may simply confirm without necessarily supplanting earlier data; studies done in 1988 don't necessarily mean more than those done in 1981. A researcher must still take care to find the most current data; a researcher in the eighties would have to justify a heavy reliance on older research. (In law courts, documents older than 20 years are considered "ancient.") Any breaks or problems in the documentation must be explained. Suppose a paper written in 1988 lists newspapers, books, and magazines for the years 1980, 1981, 1982, 1983, and 1988. Most researchers would be curious about the break between 1983 and 1988. Was nothing relevant published in those years? Did the researcher decide to skip those years? The tradition of research is that you should begin where you are and then proceed backward in a uniform, systematic fashion. (See page 40.)

Simplicity

The rule of simplicity (called "Occam's Razor") requires that we accept the simplest explanation when there are competing theories. In effect this means that there is a research bias against bizarre, exotic interpretations that require assumptions not in evidence or that require a contradiction of accepted fact. People often say that anything and everything is possible, and therefore the existence of things like UFOs, yetis, monsters, ghosts, and so on seems quite possible. But Occam's Razor requires that we not accept such things without compelling evidence. Upon finding what appears to be a huge human footprint, we shouldn't assume a monster made it, because to do so requires first the assumption that there are such things as monsters. Extraordinary explanations should be resisted until the evidence for their truth becomes undeniable.

Impartiality

The researcher must remain objective, unbiased, especially if he or she has some personal involvement in the research question. College students are expected to be quite impartial, but this can be a difficult requirement: there are subtle opportunities for biasing the data. For example, overresearching one side or underresearching the other presents an imbalance in the data which may seem natural (i.e., there really are more sources on one side than the other). However, the imbalance may simply be the result of the researcher's greater enthusiasm for one side than the other. Equal amounts of data may very well exist for both sides if the researcher looks long enough. Or—as sometimes happens—the bias may be built into the question itself: Does marijuana kill? There is so little evidence that marijuana is directly lethal that we should challenge the research question itself. On what grounds is this a worthwhile question? Its built-in bias (favoring a negative answer) will of course be substantiated by ample evidence on one side and very little on the other.

Impartiality is especially difficult when some of the evidence is personal knowledge. In school you may or may not be permitted to offer firsthand experience as evidence (check with your instructor). If you're writing about the effects of alcohol, for example, and can describe its effect on a friend (an effect you have witnessed), this is one kind of evidence. Such evidence isn't conclusive in itself, of course, but it's a useful concrete illustration to be used with other data, such as statistical studies that generalize about the effects of alcohol. The difficulty of remaining impartial then should be obvious: Firsthand observations can be highly subjective.

Common Knowledge

Common knowledge is permissible evidence; however, it isn't always clear what knowledge is common and what isn't. The general rule—that educated people by definition share a body of knowledge which is therefore "common"—isn't altogether useful. Educated people today may have widely different educations. We can take as an example the phases of the moon: new, quarter, half, full, and so on, which occur at stated times. This is "common" knowledge in the sense that it has been long known, is well established and accurately recorded in reliable reference works. However, the exact phase of the moon at any given time in history is probably not known to many people. What is or isn't common, then, is largely the result of the audience to whom you are writing and your own expertise in the matter. A renowned scholar writing to other scholars in the field may take a great deal for granted. College students writing for a general audience should take little for granted.

Relevance

There is no room for side issues. Research papers must limit themselves to just those facts that are relevant to the thesis. Once you have narrowed the thesis to the effects of smoking cigarettes, you must resist the temptation to include cigars, pipes, and chewing tobacco—enlarging the question with secondary issues. Also, if the question has been narrowed to the physical or physiological effects of smoking, you must not include data on moral, aesthetic, or other issues, even though you may have come across good material which you are reluctant to discard.

Probability

More often than not, researchers must attempt to deal with probable facts and events instead of indisputable evidence. We can't say that if you smoke cigarettes you will develop lung cancer. We can only say that the *probability* of lung cancer increases if you smoke. Nor can you as a researcher say that given all the data in my paper, it's established that President Kennedy was shot by a lone assassin. You can only say that the conclusion seems probable or reasonable given the data.

The danger here is that researchers come to associate the probable with the normal, the acceptable, and finally the expected, as when we say "It will probably rain tomorrow." Imagine a lawyer arguing that his or her client is innocent because it *is not probable* that a person of wealth, power, and prestige would have committed a petty crime. Most researchers today are cautious about what they are willing to conclude, what inferences they will draw based on probabilities. It may seem unlikely that a high government official would jeopardize his or her position for the sake of a bribe or illicit favors, but such things have happened. It is therefore not easy to argue from probability in the sense of "normal" or "to be expected."

Probability in research means a chance or trend, the statistical chance (probability) that an event will occur. It's important to know that the probability of an event is unaffected by prior events. The probability of throwing heads or tails remains exactly the same even though you may have thrown heads 100 times in a row—the probability that the next throw will be heads is no greater or less than if you had thrown no coins at all. Therefore, it's faulty analysis to assume that because a person has had one accident, he or she will "probably" have another. Nor is there any statistical probability that someone who has led an exemplary life will never be tempted to commit some crime.

Statistical Data

Researching any subject today is likely to involve the use of some statistics. Statistics can mean anything from simple numbers to complex mathematical computations. Figures can be used as incidental data, or they can be the whole body of research. The power of numbers is such that they often carry more weight than other kinds of data. Specific numbers are more convincing than generalizations about "more" or "less" of anything. Researchers who have found statistical data are often perceived as more knowledgeable and trustworthy than researchers who offer only generalizations.

In using statistics today, you must carefully consider your audience. Today everyone has calculators, everyone has had at least some high school math, and more and more people have had some computer experience; still, a surprising number of people are confused by numbers. You must try to estimate what sorts of figures your audience can handle. For example, the following statement appears in a paper arguing for a higher drinking age: "Right now, 32.4 percent of children from ages twelve to seventeen use alcohol as compared to 15 percent in 1971." Many readers will have no difficulty with the statement, but there may be some who will misread it since the figures identify a percentage of a *group* of children (not all children) and compare it with figures from an earlier time. The general reader can't handle complex figures with ease. Only a very special audience, for example, will understand statistical concepts such as the following: "With a *t* value of 8.4 and 99 degrees of freedom, we find from Table F that the mean difference is significant beyond the 0.01 level." (Popham and Sirotnik, *Educational Statistics.*)

Another problem arises when writers use too many statistics. Though numbers are powerful data, they can quickly overload the reader's mind. There is no rule about how much is too much. For example, read this sentence from a paper on the advertising industry: "These media, along with their respective percentages of the numbers of advertising dollars spent, are as follows: 30 percent for newspapers, 20 percent for television, 14 percent for direct mail, 7 percent for radio, 5 percent for magazines, and the remaining 24 percent for samples, posters, and other promotional materials." Perhaps this isn't too much for most readers, but this is only one sentence. If there are many such sentences, the reader may soon begin to nod off. If your research must make use of many statistics, consider presenting them in tables or charts (discussed in Chapter Six).

Finally, we have all learned to be suspicious of statistical generalizations. Numbers are useful and important in research, but they don't automatically solve problems. Suppose you find the statement, "There is a 50 percent chance that half of the students in any class will be wearing blue jeans." What would you be willing to conclude from this? The statement actually says only that half the students may or may not be wearing jeans. Beyond that you need to know who the students are, where the school is, what kind of classes are meant, and even what time of day and what season of the year the observation was made before you can draw any strong conclusions. People have learned to be suspicious about the sampling techniques of much modern research. Therefore, it's important to be cautious about using figures to "prove" things (see Amphiboly, p. 137). Certainly you should be on guard against sweeping generalizations, such as this one:

> It [*Saturday Night Fever*] spoke to the condition of millions of well-dressed, physically beautiful boys and girls who have everything—and nothing. Who have made love and taken drugs and lorded it over their little worlds like princes and princesses to the manner born. Modern youth are still lonely, but they're not seeking love. They have found it—in their own reflections.
>
> Albert Goldman, "The Delirium of Disco," *Life*, November 1978

Goldman may be right, of course, but we cannot tell from his statement. "Millions" is a very large number . . . millions who are all well-dressed and physically beautiful? All of modern youth? While numbers greatly add to the persuasiveness of research, careless or inappropriate use of statistical data will have just the opposite effect.

EVALUATING THE DATA: A TEST CASE

The following data represent some of the source material you might find on the question of whether apes can talk. If you use the principles of evidence already discussed, it's possible to reach some conclusions about these sources. Read the excerpts to see whether you can answer the question:

Can Apes Talk?

1 . . . nothing is more human than the speech of an individual or of a folk. Human speech, unlike the cry of an animal, does not occur as a mere element in a larger response. Only the human animal can communicate abstract ideas and converse about conditions that are contrary to fact.

Clyde Kluckhohn, *Mirror for Man*, 1949

2 Thus the basic feature of linguistic fluency is its creativity. . . . Suppose we are testing someone to determine whether or not he has gained fluency in a foreign language. Obviously we do not credit him mastery of the foreign language if he is only able to understand or produce those sentences whose meaning he has been previously taught. Analogously, we do not credit animals with fluency in language if they merely respond appropriately to verbal commands in which they have been extensively drilled. Rather, the criterion to determine if someone has acquired fluency is whether or not he can understand any sentence of the foreign language that he has not before encountered (and that a speaker of the foreign language would be able to understand). The theoretical significance of the ability to produce and understand novel sentences, then, is that this is the real test of fluency.

Jerold J. Katz, *The Philosophy of Language*, 1966

3 The large apes are our nearest relatives. This means that our genetic inheritance has developed from the same stock that produced the large apes. In our chapter on the evolution of the apes, we saw that man's common ancestor with the large apes was living, in all probability, in the Oligocene period, and that the gibbon line had already diverged from the main stock by the middle of the Oligocene; we saw too that the hominid line is not directly related to the monkey line, which was also distinct in the Oligocene. The hominid line must have become distinct toward the late Oligocene or very early Miocene, at much the same time as the forerunners of chimpanzees and gorillas were differentiating too.

Vernon Reynolds, *The Apes*, 1967

4 Communication among chimpanzees in the wild consists chiefly of "talking with facial expressions and with the hands." Holding up the hand, that official gesture of the policeman, means exactly the same thing: "Stop!" These apes likewise signal "Come here" or "Walk quickly past me," with gestures that are amazingly human. The hand outstretched in a begging gesture . . . signifies a greeting, a plea, or a recommendation that a fellow chimp calm down. The reciprocal greeting or the gesture of accord consists in holding out the hand reversed, that is, with the palm down. In such a gesture the fingertips of the two animals may touch. But the gesture can also be well understood at a distance.

Vitus B. Droscher, *The Friendly Beast*, 1970

5 The two chimpanzees that are learning "language" have done so by different systems, which are intended to reveal rather different aspects of language. Premack (1970) taught a young female chimpanzee named Sarah to place variously shaped pieces of plastic on a magnetized board. Each plastic chip represented a word, whereas a string of chips was a phrase. . . .

Within these limits, Sarah clearly used symbols. She had about forty words: "same" and "different"; "yes" and "no"; "on," "under," and "insert"; and a number of nouns and adjectives. As Premack said, he was not concerned in ascertaining the possible size of a chimpanzee's vocabulary, but its possible grammatical complexity. He tested Sarah to see if she was really using the words as symbols by asking "Apple same as . . . ?" and then offering a number of choices: red or green, round or square, and so forth. Sarah's "word" for apple was a blue triangle, but she described it as red and round, as with a stem, and less desirable than grapes.

Alison Jolly, *The Evolution of Primate Behavior*, 1972

6 Washoe, on the other hand, learned a far more open system, which raises all the ambiguities and questions that surround language-learning by human children. The Gardners taught Washoe the American Sign Language for the Deaf (ASL). . . . ASL is not finger spelling. Each position of the hands corresponds to a separate word. The Gardners call her words "signs," which would be confusing in the present context; so, except in direct quotes, I shall call them words. ASL, as the Gardners say, "is a language by the most widely used criterion we have: that it is used as such by a community of people."

Jolly, 1972

7 The Institute for Primate Studies now has about a dozen chimps with varying degrees of skill in using Ameslan. Fouts has now begun to extend and fill out the original work done with Washoe. One would like to think that criticisms of Washoe would provide some points of departure for his work, that the critics would have used Washoe to focus attention on some kernel of linguistic ability that Fouts might then seek to explore. Unfortunately, the critics have focused their attention on the deep anxieties summoned by the idea that a chimpanzee might be capable of language. And so, essentially, Fouts has had to start afresh,

using Washoe as the bedrock on which to construct a new view of language, rather than using her to modify old views.

Eugene Linden, *Apes, Men, and Language*, 1974

8 When conversing with Roger in Ameslan, Lucy would look at him with intense concentration; however, her movements in making signs were not intense, but leisurely, as though communicating by using Ameslan was the most natural thing in the world for a chimp to do. She seemed to understand spoken English. It was eerie to be talking with Roger about Lucy's mirror or doll and then have her run over and pick it up. Roger noted that earlier that week he had lost Lucy's doll. He glossed his error by replacing the doll with a slightly different one, which he handed Lucy the next day. Lucy was very suspicious of this new doll, and the day after this surreptitious exchange she went over to her toy chest and signed to Roger "out baby." She wanted to see where this strange doll had come from.

Linden, 1974

9 For instance, if a chimp signs *come-gimme tickle* as frequently as he signs *tickle come-gimme*, it may simply be cranking out signs appropriate for the incentive of being tickled, which is not the same as generating a sentence. And, Brown argued, since word order is "as natural to a child as nut gathering is to a squirrel," unless one had frequency data for appropriate and inappropriate word order, Washoe's "semantic intentions" would remain a matter of guesswork.

Stuart Bauer, "First Message from the Planet of the Apes,"
New York Magazine, 1975

10 What the apes appear to have learned is not the meanings or grammatical functions of signs, or how to combine them using productive rules, or how to decode what was signed to them, but rather that signing behavior was very important. The activity was highly valued by their teachers. Under these circumstances, they learned that the mere behavior of producing signs could be used to effect certain outcomes—e.g. getting food, social approval . . . etc.

Mark S. Seidenberg, and Laura A. Petitto. "Ape Signing: Problems of Method and Interpretation." In *The Clever Hans Phenomenon*, 1981

Considering this evidence, what conclusions can be reached? You must evaluate all the evidence, individually and collectively, before you can begin to answer, even for yourself, the question of whether apes can talk. About this evidence you can say the following:

Item 1 It would be nice if all the experts agreed on your research question, but probably they won't. Kluckhohn's statement that only the human animal can communicate abstract ideas seems to contradict the idea that apes can talk. What to do with Kluckhohn's statement? A researcher must account for contradictory data; you can't simply discard Kluckhohn or forget that you've read him.

Don't undertake research to prove a preconceived idea: don't commit the *a priori* mistake. A single piece of data that doesn't fit may be more important than all the others. It would be a mistake to start out to prove that apes *can* talk. If you phrase your research question so that either answer is appropriate, you won't have to ignore contradictory information. The question should be worded, "Can apes talk?—an investigation into research on both sides of this question." With that qualification, Kluckhohn is no longer contradictory; he is simply one of those who may think apes can't talk.

Note the date of Kluckhohn's book. It wouldn't be wise to draw any strong conclusion from Kluckhohn's statement. Later research may have made him irrelevant. Kluckhohn himself may have changed his mind since this was published.

The biggest question to answer is, So what? Kluckhohn doesn't directly address the matter of whether apes can talk. Perhaps he isn't even relevant. So what if Kluckhohn thinks human speech is "unlike the cry of an animal"? You can't determine the significance of Kluckhohn's statement until you know the significance of your own question. Why is it important to find out whether apes can talk? Is this merely an amusing question? Is the question merely curious? Or is there some deeper significance? While Kluckhohn's book doesn't seem to address itself to the practical side of the question (data on apes), does it offer anything on the theoretical side (significance of the study)? See the Linden note (item 7) about "deep anxieties summoned by the idea that a chimpanzee might be capable of language"; there may be more to Kluckhohn's study than you at first imagined.

Item 2 Kluckhohn is an anthropologist; Katz is a linguist: Neither of them seems to be talking specifically about apes. Other than the theoretical implications, are their statements relevant to the question? Kluckhohn and Katz suggest that there is a very important qualification to the question: namely, it depends on what you mean by "language," what you are willing to accept as "talk." It will do no good to say everyone knows what talking means. Nor will it do any good to look it up in the dictionary. Two scholars in the field you are investigating suggest that you must define your terms before you can answer the question.

Kluckhohn suggests that talking is communication of "abstract ideas" and conversing about "conditions that are contrary to fact." Katz suggests that talking is the ability to "produce and understand novel sentences." On these "tests" of language, both of them seem to doubt that any animal can "talk." Therefore, the previously simple question—Can apes talk?—has now become complex. There is no way to get around the complication. Your report will have to examine carefully what is meant by "talking" or "communicating" from anyone who would suggest that apes can talk.

Item 3 The Reynolds quote is interesting because it seems not at all relevant. Since it's painful to discard anything once you have gone through the process of finding and copying, the temptation will be strong to find a way to use it. Perhaps the Reynolds quote can be worked into the introduction of the paper as "background"? Perhaps it can be worked into the conclusion as you speculate on the relationship of human beings to their ancestral brethren? Both of these are poor ideas. They require you to distort the data or to distort your study. In *your* study, the evolution of the hominids isn't relevant. The Reynolds quote should be discarded.

Item 4 Droscher's book seems to be the first one to concern itself clearly and directly with the question of whether apes can "talk." But it isn't clear what we should conclude from this. He tells us that apes do communicate with each other by means of gesture and facial expressions. Does this fit your question? Maybe it does. But you will further have to qualify your definition of "talking" to include sign language and nonverbal communication. If you do so, you will raise an important (new) question—are you getting so far removed from the normal definition of "language" that no one will be able to agree with you? Will other researchers say, "Well, of course, if you mean something like nonverbal *signs*, even my dog can 'talk.' He barks when he wants to go out, looks quizzical when I talk to him, and so forth. But that is *not* the same as human use of language." Certainly Kluckhohn and Katz would say that kind of behavior doesn't fit their definitions for human communications. So you must think carefully about Droscher's contribution. At least, it's safe to conclude, certain kinds of communication seem within the apes' capability. Droscher seems to be saying something important, but you should examine the rest of your data before deciding what.

Items 5 and 6 Jolly's contributions get to the heart of the matter. She describes the "languages" being taught to apes—one an abstract symbol system, the other a sign language. Now you can better see the relevance of all the previous research. The "tests" of language suggested by Katz and Kluckhohn can be applied against these languages. And now, for the first time, there is evidence that, indeed, apes can talk. Certainly Sarah and Washoe seem to be using language within the normal definition of the word. Droscher's statements about sign languages in the wild now seem to relate to the experiments with Washoe, who was taught American Sign Language. You could conclude that although speech isn't natural to animals, sign language is. Therefore it isn't surprising to discover that animals can talk when provided with a language suitable to them. However, a little additional thought may raise another troublesome notion. The very fact that sign language is natural for apes could mean that Washoe and the others have simply been coached to make appropriate signs on command, like the trained dogs and seals in the circus. So the issue is not settled, even though Jolly's data are very interesting.

Items 7 and 8 Linden seems to be backing up Jolly, but here it's important to understand the source of your material, especially to know who your authorities are. Alison Jolly is a scientist interested in animal behavior. For some of her data, she relies on other research: she cites Premack and the Gardners. Linden is a free-lance writer who also relies on other research: the Fouts, Premack, and the Gardners. Both Jolly and Linden are respected writers. However, although they seem to be confirming each other's work, they are in fact relying on the same information in some cases. They are both drawing conclusions based on primary research done by someone else. Thus, instead of two sets of data, you really have only two views of one set of data. This doesn't diminish the contributions of Jolly or Linden. It merely suggests that before you draw any hard conclusions from their data, you should look at the work they cite—Fouts, Gardners, Premack—listed in the footnotes. (Here you have evidence of the importance of reference notes and bibliographies; serious researchers will help you find the material you need to evaluate their contributions.)

Item 9 You could, of course, content yourself with Linden and Jolly. They are both well-informed writers and, you may think, far more knowledgeable than you are. But one of the virtues of research is its total impartiality. You don't have to be an expert to make important discoveries if you're careful and thorough. Furthermore, time marches on—and with time may come new discoveries. Although Linden's book is dated 1974, the research he relies on is considerably older than that. Baur reports more recent research in his 1975 article. Hence, it would seem, an important criticism is being raised about sign language. Perhaps those who claim that apes can talk are wrong after all. How important do you think Brown's argument is? Do you need to see Brown's work, or are you satisfied with Baur's description of it?

Item 10 The last item is the most recent. For that reason it may seem to be the most important. Seidenberg and Petitto are suggesting that apes do not really understand language, they simply crank out arm waves or finger wiggles to get what they want. Thus it might appear that the most recent research is negative. These authors raise a significant objection that would affect the interpretation of all ape-language research, if they are right. But you cannot tell from this excerpt whether Seidenberg and Petitto are right; you need to see more of their research, the evidence on which their conclusion is based.

Summing Up the Evidence

It may seem to you that there is no end to this question; each new piece of research just raises new questions, and perhaps you'll never be able to answer the original question. All this is true to a degree. To be a researcher means to give up your old way of thinking. Very little in life can be answered in the simple yes/no, right/wrong terms used with children. Even presumably elementary and obvious questions will, on examination, turn out to be quite complex. It shouldn't be surprising that the more you research, the more questions you will raise. It's true that for your most serious inquiries you usually will not be able to find comforting, absolute answers.

Because of the inconclusiveness of most data, many researchers are very careful about the kinds of conclusions they are willing to draw. If the data presented here were all the data available, what would you be willing to conclude? You should assume that all these researchers are correct. (The possibility that one or more of them may be incorrect exists, of course, and if proved would seriously complicate your research, but unless the error is obvious and presents itself to you, there's no need to go looking for it.)

If the question is,"What does the research show about the question of whether apes can talk?" and this were all the research you had, it wouldn't be safe to conclude that apes can talk. You could say there is some evidence suggesting that apes may be able to talk, depending on how terms are defined, but the question is still open to debate. Some researchers aren't entirely satisfied with the definitions of "language" being used. Moreover, Sarah and Washoe may be freaks. Most researchers would hesitate to answer the question on the basis of what two chimpanzees are reported to be able to do. (Can these results be duplicated by other researchers using other chimpanzees?)

On the other hand, if you assume Jolly and Linden are correct, or at least not incorrect, the evidence on Sarah and Washoe won't simply go away. It's too easy to say the case isn't proved and so should be dismissed. That is, it wouldn't be good to conclude from this data that apes can't talk. Have you made any progress on this question at all? Yes, you have. The present research, while inconclusive, demonstrates that this is a serious research question. Whereas we may once have been able to laugh the question away, it now appears that serious doubts have been raised on both sides of the question. Disturbing new questions are causing researchers to rethink old positions, and researchers are beginning to see what data are needed in order to answer the question satisfactorily. Thus, it's appropriate to say that further research is necessary. So many studies end with a call for more research that it's a research cliché. However, you have from Baur's article a specific suggestion about the nature of the future research. See the Appendix for a research paper on this question.

FALLACIES: PROBLEMS WITH EVIDENCE

Assumptions are beliefs about prior conditions that must have existed (we believe) in order for present conditions to exist. If we find the charred remains of a building, we assume there must have been a fire. We could be wrong, of course—someone may have deliberately contrived the evidence to mislead us. We also assume people who are now in jail must have committed some crime, but this assumption is quite a bit more shaky since it rests on the still earlier assumption that anyone in jail must be guilty. The statement overlooks the possibility of the innocent person wrongly convicted.

Assumption, inference, and *premise* are used more or less interchangeably. However, they are not exact synonyms.

- An *inference* is a deduction, a conclusion we reach about data. It is used chiefly in contrast with *factual* data (i.e., factual contrasts with inferential data).
- *Assumptions* are generally antecedent beliefs not in evidence. We *infer* that a weight will fall when dropped because of the data before our eyes. We see that the weight appears heavier than air (and will not float) and will respond to the force of gravity (will not be artificially held aloft when released). These inferences, we may say, are based on prior *assumptions*—that heavy objects fall, that gravity exists. If for some reason the weight should surprise us and not fall, we would immediately suspect that our *inferences* about the physical evidence before us were wrong. We would resist challenging our *assumptions* (our basic beliefs) until we were thoroughly satisfied that our inferences were correct. That is, only if we could be certain that the weight was in fact heavy and gravity was operating on it would we begin to suspect that one of our assumptions was inaccurate.
- *Premises* are assumptions, but in research the term "premise" is usually reserved for those assumptions on which experiments are constructed or on which reasoning depends. Premises are the necessary assumptions in a given case. They are those assumptions or inferences we seek to test, to establish, to corroborate, or to discredit. Today the word *premise* is frequently replaced with *hypothesis*. In a given case you may find it helpful to use all three terms in the analysis of evidence. You can speak of a researcher's premises, meaning the necessary or immediate beliefs inherent in the research—the hypotheses. Or you may speak of the researcher's inferences, meaning the conclusions drawn from the data. And you may speak of the researcher's assumptions, meaning fundamental, underlying beliefs.

Because assumptions are beliefs rather than facts, we can't verify them; we can only reason about them. It is here that research is most vulnerable to error. Authorities you quote can be as susceptible to error as you are. Faulty assumptions can underlie articles you find in magazines and newspapers. The most frequent kinds of errors have to do with the implications of data—the conclusions of research. An implication is a necessary consequence of prior conditions. If we find the suspect with a smoking pistol in his or her hand, standing over the victim, the implication is that the suspect is the killer. But jumping to conclusions is the opposite of research. You must be very cautious about the kinds of conclusions you are willing to accept from authorities and the kinds of conclusions you will permit yourself to make. The problems inherent in reaching conclusions are commonly referred to as the *fallacies of evidence.*

ACTIVITY 26 Determine the hidden premise or assumptions or the implied assertions in the following statements. Write out the implications (there may be more than one).

1. Imprisonment is at least partially effective as a means of reformation because 20 to 40 percent of offenders never return to prison after being released.
2. Our enemies only understand force.
3. It's nonsense to assume Donny died from drinking the iodine since it is clearly labeled "Poison."
4. A young girl taking the pill had trouble with vomiting, breathing, and a mysterious pain. She soon died.
5. Young men of America must show that they are made of the same stuff as their loyal forefathers.
6. If we don't want a nuclear war, we must convince the rest of the world that a holocaust will follow any attack on us.
7. It's impossible to know whether aid to developing nations ever reaches the people for whom it is intended.
8. The trouble with kids today is that they lack respect for their elders.
9. People who can't swim should stay away from Lake Michigan.
10. Finding the answer to the true cause of alcoholism lies in the road ahead.
11. If we don't protect Europe, our allies will be enslaved.
12. If the Soviets have twice as many missiles as we do, it means we are only half as well prepared as they are.
13. He was too young to die.
14. God doesn't intend for this planet to fall under the domination of atheistic communism.
15. We must rededicate ourselves to decency and integrity.

Problems of Insufficient Evidence

Overgeneralizing

The primary judgment for any researcher is whether there is enough evidence. Thus the chief problem is overgeneralizing, reaching conclusions too quickly on the basis of too little evidence. How many accidents does it take to convince us that nuclear plants are unsafe? Much depends on the quality of the evidence, as well as the amount. One hundred minor accidents may not be as persuasive as one major incident. You must weigh pluses and minuses and not form snap judgments on simple numerical evidence; potential risk may be as important as incidents that have already occurred. It may take much sorting of testimony and data to identify real physical risks, intangible (psychological) risks, acceptable and unacceptable risks. Letting the reader see your reasoning (not just your conclusion) is your best insurance against hasty generalizing.

The common notion that a single exception disproves any generalization isn't always true; some of our generalizations achieve the status of law or general truth, even if we can find an exception now and again. We have a general truth that smoking cigarettes greatly increases the likelihood of cancer, the occasional cancer-free smoker notwithstanding. If we know that lead will melt at relatively low temperatures, we won't reject this rule just because in some experiment lead has resisted very high temperatures. We are more likely to doubt the experiment or to suspect that the lead is really some kind of alloy. While it's true that you must be wary of hasty generalizations, there is an additional caution not to dismiss every general law on the grounds of possible exceptions. We require a great deal of evidence to overthrow established truth.

Card Stacking

Ignoring some of the evidence is of course unscientific and in some cases can be considered fraudulent. This is card stacking. You must not select data in order to favor one side of your research question. You can avoid this possibility in the first place by avoiding research questions in which you have a strong personal involvement. Researchers who use drugs aren't likely to be able to remain objective in gathering data against drug use. No matter how careful, fair, and impartial it is, such research will be suspect. If you have a personal involvement, you may overlook data, or you may less enthusiastically pursue data that disappoint you.

The other remedy against card stacking is to frame your question in such a way that either answer will be acceptable. Instead of setting out to prove *a priori* that marijuana is harmless, ask "What does science know about marijuana?" The first question is biased and suggests a writer who merely wants to argue in favor of his or her own point of view. The second question implies no bias and allows the researcher to gather data on both sides of the question.

Ad Ignorantium

Arguing on the basis of a negative, or on the basis of what is not known, is the appeal to ignorance, the fallacy of *ad ignorantium*. It is presumed that if you can't prove something is false, it must be true. That is, if we can't prove that there is *not* a monster in Loch Ness, it must be (or may be) true that the monster is there. The same problem is involved whether you insist on the validity of anything not proved false or the falseness of anything not proved true. The fact that we don't know something or cannot prove something is no grounds for any kind of deduction. Students who are working with supernatural subjects (ghosts, witches), bizarre or unusual subjects (the abominable snowman, Bigfoot, the Bermuda triangle), or extraterrestrial phenomena (Martians, UFOs) must be careful to avoid falling into the *ad ignorantium* trap. (See the Rule of Simplicity: Occam's Razor, p. 117.) In everyday conversation we may say that anything is possible, but in research it's necessary to maintain a healthy degree of skepticism about any conclusions that are offered merely because there is no evidence against them.

Post Hoc Ergo Propter Hoc

Literally, *post hoc ergo propter hoc* means "after this, therefore because of this." It refers to cause-and-effect events. The fact that one event follows another in time doesn't mean the events have a cause-and-effect relationship to each other. They *may* have such a relationship, but you need more information to establish cause and effect. This can be a very difficult judgment to reach because any number of events may happen together. There may be a chain of events, only some of which are visible to the researcher. If we light a match near an explosive substance, the resulting explosion may or may not have been caused by the match. Then, too, matters can get complex when you have to determine immediate causes and prior or contributory causes. The accident was immediately caused because your brakes failed . . . but you knew the brakes were going, so a contributory cause was your negligence, and so on. The whole business of causality can be so complex that some researchers avoid it entirely. In any case, it's wise to be cautious about attributing cause and effect on the basis of the sequence of events in time. It may rain after we do a rain dance, and conceivably our dance may have been the cause of the rain; but we have no way to know that—there is no necessary connection between the two events.

Problems Based on Irrelevant Information

Ad Baculum

The Latin word *baculum* refers to a stick or club; thus an *argumentum ad baculum* is an appeal to force. Appeals to force, extortions, or threats, while difficult to ignore, obviously have nothing to do with the merits of the evidence. Interests other than pure science have sometimes influenced the results of military and industrial research. In research on the effects of smoking tobacco, to pick just one example, tobacco company researchers sometimes found that smoking tobacco didn't cause cancer. Such research may be valid, but the suspicion is present that industry interests may have had something to do with the results. Even "pure" researchers faced with budget cuts, loss of programs, and so on can be subject to pressures that may influence their results.

Ad Hominem

Ad hominem means "to the person." *Ad hominem* judgments usually ignore the relevant data entirely and instead focus on some personal fault. Critics writing about performing artists (actors, dancers, singers) are sometimes unable to deal fairly with the performance because of some personal quality of the performer. Those who live unconventional private lives or who hold controversial moral, social, or political views may find that their views are attacked even when they are irrelevant to the issue. In secondary research you must be sure that the authorities you are reading are talking about the relevant subject matter and not merely indulging in name-calling and *ad hominem* attacks on other researchers: for example, "No one takes Smith's research seriously; his communist sympathies are well known." It's better to look at the evidence itself and make your judgments on its merits rather than on the personal qualities of the researcher.

Genetic Fallacy, Fallacy of Opposition

Related to the *ad hominem* fallacy are the *genetic fallacy* and the *fallacy of opposition* (the terms are often used interchangeably). In both cases, the evidence itself is suspect because of its origin. Western medicine, for example, long resisted acupuncture as "merely Chinese folk medicine." That is, the fact that acupuncture came from the East influenced its evaluation in Western medicine. Then, too, it's common to suppose that great institutions must produce more reliable research than do smaller or less-well-known ones. If the University of Chicago or some other great school announces something, many researchers will treat the announcement as definitive because it comes from such a "reliable source." If the same announcement came from Montezuma State College or Mickey Mouse University, it would not be given the same credence. It isn't a good idea either to accept or reject evidence solely on the grounds of where it comes from. Great authorities and institutions can be wrong, and quite humble institutions and researchers may do significant research. The best approach is to examine the data directly and not get sidetracked by irrelevant issues.

Guilt by Association

The "guilt by association" fallacy assumes you are like your friends, that we can judge your ideas not by evaluating the evidence but by looking at those you associate with. In theory, you should be able to associate with people without sharing all their beliefs, values, or behaviors. A researcher should be able to study communism without becoming a communist. A lawyer should be able to defend criminals without becoming one him- or herself. Writers who write intellectual essays should be able to explore controversial subjects without losing credibility. We may find it necessary to defend the rights of murderers for the sake of our principles of civil liberty, and this shouldn't mean that we are fond of murder. However, in practice, at least in the public mind, it's very difficult to separate yourself from your associates.

Ad Misericordiam

Ad misericordiam is the appeal to pity. For example: "We should all contribute to charity to aid the misfortunate," or "Soldiers wounded in the service of their country should be rewarded." We can't say that appeals to pity are automatically irrelevant; obviously it is relevant information that some people are in need of aid. However, these appeals are especially susceptible to manipulation and can be a form of extortion—psychological blackmail. Therefore we must make sure that we are well informed when faced with *ad misericordiam* appeals. Is the claim legitimate (are people suffering?)? Will funds raised reach the intended recipients? We can't afford to be naive about such appeals. History has shown that some people have managed to make a profit out of charity. It's important also to remember that your emotions can quickly overwhelm your ability to analyze the matters at hand. Tiny aborted fetuses, pictures of mangled bodies, and other gruesome evidence can be powerful emotional appeals. Is such evidence unfair? It depends on the case.

Ad Populum

Ad populum means "to the people," but the sense of it is "to the mob": appeal to popular prejudices and slogans. "Down with big government!" "Down with invasion of privacy!" "Vote for law and order!" "America for Americans!" A populist is someone running for office by appealing to popular or traditional ideas, campaigning against social change or newer attitudes. There is nothing wrong with clinging to traditional values, but populism usually insists that such values are superior simply because they are traditional. ("If it was good enough for Abe Lincoln, it's good enough for you.")

The most objectionable aspect of populism is its sloganeering. You should remember that when rascals seek to harm you, they will often cloak themselves in virtue and give some high-sounding name to their action like "patriotism," "love of country," "the American way of life," "decency," and so on. Thus populist slogans often become code words for injury. During the 1960s the call for "law and order" became a code word for discrimination against racial minorities, anti-war demonstrators, and anyone else thought to be troublesome.

Ad Verecundiam

Ad verecundiam refers to inappropriate authority. The chief reason for examining the credentials of authorities is to discover that some so-called authorities aren't really experts at all. When celebrities, important people, or any "experts" are cited, you must be sure to investigate their expertise. Prominent people are frequently asked their opinions on matters outside their area of expertise. Athletes and performing artists are often asked to "testify" (sell) in behalf of products that have nothing to do with athletics or art—products on which the speaker has no more authority than anyone else. Even scholars and scientists are sometimes asked to offer an opinion on matters in which they are not expert. Such people have a right to give opinions, of course, but this material should be seen for what it is. No strong conclusions can be based on such evidence. The problem can become especially complex when one authority relies on another, and for that reason it's important for you to try to find out who your authorities are and what their expertise is.

Bandwagon

The bandwagon approach is an appeal to peer pressure, to group identity. This appeal is used heavily in advertising: 5 million Americans read X *magazine* (its popularity must mean that there is something good in it—you will probably enjoy it too). The bandwagon appeal can be aimed at "plain folks"—those who think of themselves as ordinary, or at snobs—those who think of themselves as superior.

Researchers sometimes can be intimidated by a bandwagon effect from the weight of research itself. Since the majority, established position in your area of study is X, you would be going against the majority to argue otherwise. However, the impartiality of research is such that the majority opinion can be wrong too. (See also Genetic Fallacy, p. 132.)

Common Sense

Another fallacy is based on an appeal to common knowledge or practical truths: "The world cannot be round like a ball; otherwise the people on the bottom would fall off," or "obviously nothing heavier than air can fly." Appeals to common sense aren't always wrong, but they are often used to simplify—or oversimplify—difficult issues or to assert the virtues of the commonplace over intellectual or unusual values ("I may not have studied art, but common sense will tell you modern art looks like cartoons, children's art"). Common sense isn't helpful in matters requiring *uncommon sense.* Frequently our old assumptions, our accepted definitions, and our practical view of the world and reality get in the way of research. It's better for researchers to make no assumptions whatsoever about data. Our "common-sense" perceptions, in fact, can become mindsets that prevent us from finding new ways to look at things, or even sometimes prevent us from seeing what is obviously right in front of us.

Red Herring, Straw Man

Arguing beside the point, switching to some side issue or an entirely new issue to distract from the main argument, is a favorite device of people who have been asked hard questions. For example, imagine a member of congress, when asked whether busing is necessary to achieve integration in schools, giving this answer: "There are so many difficult issues connected with busing. Have you seen those buses? Some of them are twenty years old, rickety old death traps. I tell you I wouldn't want my child carried halfway across town in one of those old crates."

The answer here seems to be about bus safety. Our hypothetical politician seems to have said "no" to busing but has sidestepped the issue of integration entirely. The name "red herring" allegedly refers to escaping prisoners' practice of drawing a fish across their trail to confuse the bloodhounds and make the dogs lose the main trail. The technique is well known to students faced with exam questions. For example: "In order to understand the causes of the Civil War, we must look at the warlike nature of human beings. The history of Europe is full of wars. . . ." Here the student has worded the answer to a question about the Civil War so that he or she will be able to discuss war in Europe. It isn't a good technique.

The "straw man" argument is related to the red herring. The straw man is invented so that it can be knocked down. For example, in an argument about whether we can afford a National Research Institute, imagine the opponents inventing someone (a "straw man") called "the researcher" or "researchers today." This straw man is a generalization that treats all researchers as if they were the same. But it's a very convenient invention; it permits the opponents to argue against the straw man instead of against the real issue. "Researchers today," we may be told, "are just a bunch of freeloaders; they get their hands on big government grants for the most ridiculous research projects." The straw man, like the red herring, permits the writer to avoid the main question and deal with some side issue.

These examples may be too obvious to be useful. In actual cases the side issue may be much more difficult to separate from the main issue. Suppose you were to investigate the question of whether subliminal advertising works. Messages that appear and disappear too quickly to register on the conscious mind or are hidden among other information so that they register only subconsciously are *subliminal* (below consciousness). The question is, do they work? Can subliminal popcorn ads at movie theaters cause you to buy popcorn? The answer to the question requires data on sales with and without subliminal ads. Note that nothing is said about the ethics of this kind of advertising; the question isn't whether advertisers should be allowed to use such a device. The question only asks whether it works. However, the two issues are closely related, and a student who sets out to examine one will almost certainly come across the other. Therefore it's possible to get the two things mixed up in the same paper and so lead off (red herring) the reader from the main issue.

Consider an even more difficult problem. Suppose in the subliminal research you acquire some data on subconscious stimuli. Pretty girls and handsome athletes are used to sell cars and other products because these attractive people set off subconscious responses in the viewers, and that sounds like it might belong in the paper on subliminal advertising. There is an important difference here, though. The models and athletes are clearly visible; we may even realize that their purpose in the ads is to associate their desirable physical qualities with the product being sold. On the other hand, subliminal cues aren't visible. If the cue is audio, it will be below the register of our ears. If the cue is video, it will be too fast or too disguised for perception by our eyes (at least it won't register as visible). Thus there is an important difference between subliminal cues that we can't see and can't know are operating, versus subconscious responses to clearly visible or audible stimuli.

Tu Quoque

Tu quoque (pronounced "too-KWO-kway") means "You did it too." The idea is that you can't accuse anyone of a "sin" that you have also committed (if it was all right for you, it should be all right for me). By this reasoning, criminals in prison can't pass judgment on criminal matters since they themselves have committed crimes. (Yet criminals are sometimes very astute on such matters precisely because of their personal experience. Ex-criminals sometimes make excellent employees in anti-crime businesses.) One researcher may claim that his use of a faulty procedure should be blameless since he borrowed it from the works of experts X, Y, and Z. Commonly, we all try to excuse petty misconduct on the grounds that it is widely practiced ("Everyone does it"). However, that isn't relevant. No matter who is doing the accusing, the only relevant answer is whether the charge is valid. Mistakes and "sins" are wrong even if others are guilty of them. The Nuremberg trials after World War II established that you are required to break laws rather than commit evil in the name of the law. The fact that "everybody" is doing something doesn't make it right.

Problems Based on Ambiguity

Amphiboly

Amphiboly (am-FI-bo-ly) refers to language ambiguity, a deliberate or accidental misuse of implications. "Three out of four doctors recommend this type of pain relief." The implication here is that "three out of four" means 75 percent of all doctors and also that what is true of a type of pain reliever is true of this one. It's highly unlikely that anyone could have interviewed all doctors; it's possible that only a very few doctors (possibly only four) were interviewed.

Begging the Question

With circular reasoning, or begging the question, the conclusion is simply a restatement of one of the premises. "The President is such a good man . . . because he is so moral!" "Murderers should be executed because they are killers!" "You can see that God exists . . . because of all the things that He made!" In general use, begging the question or giving tautological answers means giving answers that are nonresponsive. Upon examination it usually turns out that such answers have merely turned the question around. For example, a difficult question for science and religion is, "How can we tell when a person is dead?" There are a number of nonresponsive answers we might give: A person is dead when he or she is no longer alive, when the cause of death is lethal, when the death certificate is signed, when there is no sign of life. None of these answers is helpful; we don't know any more than we knew before—we have to ask when is the person no longer alive, when is the cause of death lethal, when do we know it's time to sign the death certificate, which are the signs of life?

A classic example shows that it's possible for the question itself to be tautological: Can God make a stone so heavy that even He cannot lift it? Instead of indulging in pointless circular attempts to answer this question, it's better to analyze the question itself for circularity. The question seems like an unanswerable puzzle on the assumption that God can do anything. If this is true, the question is tautological: It asks whether God can do something that God cannot do.

Equivocation

Equivocation (e-kwi-vo-KA-tion) is arguing over the meaning of a word, deliberate or accidental misuse of the connotations of a word to disprove or distort an argument. "Senator Gonzo claims to be a conservative, yet he lives lavishly!" Two meanings for *conservative* are involved here. Politically, *conservative* needn't have anything to do with one's personal lifestyle. Another example is "How can you claim to be pious? You never go to church!" The word *pious* does mean religious, but it's a distortion to suggest that only those who go to church can be pious.

False Metaphor or False Analogy

Sometimes a metaphor or analogy has more dissimilarities than similarities. "The President has seen us through a crisis of state; he has kept late hours in lonely vigil; he has brought us soothing relief; he has wasted his strength to revive our faltering nation. Let us reward him now with our gratitude!" The metaphor or analogy of the President as doctor tending a sick nation is false, however poetic it may be. Nothing can be proved by analogy alone, and it's often said by researchers that all analogies are false and should be avoided.

Suppose we would like to argue that a current situation is like a previous situation—our involvement in Latin America, for example, has been said to be similar to our involvement in Vietnam. Since the previous situation turned out to be a disaster for us, the later situation may prove disastrous as well. (We are supposed to learn from the lessons of history.) Only if the two situations are truly alike can we reach the conclusion that the outcome may be the same. It's conceivable, of course, that some situations may indeed be very similar. In that case we would be wise to be cautious. More often than not, however, the situations aren't similar; the comparison isn't good; the analogy doesn't hold.

Most researchers advise that you treat comparisons with skepticism, especially poetic (metaphoric) comparisons. We speak of the nation's affairs as the "ship of state," and doing so allows us to conjure up all sorts of seagoing images—rough waters, foundering on rocks, smooth sailing, hand on the rudder, steering a straight course, and so on. At least when presented with analogies and metaphors, you should insist on making an analysis of the points of comparison to see whether the analogy is valid. Given the differences in time, place, and people involved, you will need a great deal of information to show that any two situations are similar.

Problems Based on Faulty Logic

Loaded Questions and Issues

Loaded questions are not safe to answer because any answer will put you in an unfavorable light. The classic example—"When did you stop beating your spouse?"—seems to allow you only two (bad) answers: "I haven't stopped" or "I stopped some time ago." In either case the question itself suggests that you are a spouse beater, regardless of your answer.

The loaded question can be worded as a loaded issue: "Let us examine whether the President's aggressive foreign posture is weakening our bargaining power!" It's necessary to determine first whether the President's so-called posture is aggressive before we can examine its influence on our bargaining power.

False Dilemma

Either/or thinking, the false dilemma presents only two options, both usually unattractive. "Either we must support private charities or we must increase welfare taxation!" "If you don't quit smoking, you'll die of lung cancer!" The dilemma is false whenever there are other, unmentioned options or possibilities. In almost any situation there are likely to be more than two options. The old choice offered to most of us was "Either pay your taxes or go to jail." In fact, it turns out that there are sometimes other options, one of which is to have Congress excuse your taxes. Dilemmas can occasionally be valid, but since they usually aren't, it's necessary for you to analyze the options.

Non Sequitur

Literally meaning "it does not follow," the general term *non sequitur* (non-SE-kwi-ter) is applied to any fallacy in which the argument can't be followed. A *non sequitur* argument is one whose conclusion doesn't follow from the premises. "Inflation has made our money worth less, so we might as well spend it and enjoy life!" "She hasn't found any significant results in any of her latest experiments; she must be incompetent!" *Non sequiturs* usually involve leaping to conclusions.

Rationalization

Rationalization is making excuses, choosing the least threatening or the most self-serving explanations. "I'm flunking calculus; that instructor hates me!" "Of course I hit the mailbox; you've got the stupid thing where it's impossible to miss!"

Reductio ad Absurdum

Reductio ad absurdum, "reduce to an absurdity," is disproof of an argument by showing some absurdity to which it leads if carried to its logical end. Frequently the user is merely being sarcastic or erroneously leading to an absurdity that doesn't necessarily follow. For example: "So monkeys can talk, you say; the next thing you know we shall have an orangutan for President!" The speaker has suggested that if monkeys can talk, then they must be human. Since running an ape for President is clearly ridiculous, the speaker hopes to ridicule the first idea itself—that apes might be able to talk. Another example: "So you would give the government the power to tax, would you? The next thing you know they'll use that power to take away your home and property, and finally your life. We will then have the amusing consequence of the government taxing its citizens out of existence—and finally the government itself must fall with no citizens and no new taxes to support it!" Some economists today might argue that this is a legitimate *reductio ad absurdum*, that governments can and do use taxation to destroy nations. Others would argue that taxation is a necessary evil and that only the abuse of taxation powers is likely to lead to collapse of the economy.

Slippery Slope

One thing leads to another. The slippery slope argument seems particularly persuasive when turned against anything immoral, illicit, or unwise. "If you eat desserts, you'll end up weighing three hundred pounds!" "If you take one drink of wine, you'll take another, and soon you will be an alcoholic!" Of course it's possible that one thing *may* lead to another, but it obviously isn't true that one thing *must* lead to another. The effect of slippery slope reasoning, if it were universally applied, would be to enshrine the status quo for fear of all innovations or changes. There is a strong fear, for example, that euthanasia will lead to a general disregard for life and may lead from there to mercy killing for the mentally impaired, physically handicapped, and so on until it becomes common to get rid of any unwanted or inconvenient people, including the poor and aged. The fear is real and not to be dismissed lightly. Yet we must have some means for change, innovation, and experimentation or the whole concept of research is undermined. Not everyone who drinks becomes an alcoholic. Not everyone who experiments with lifestyles rejects tradition. We may wish to rid society of things like hard drugs, alcohol, tobacco, and so on for the harm they do, but we must be careful not to overstate the case. Exaggeration undermines the credibility of the researcher.

ACTIVITY 27 Identify the fallacies in the following statements. Some statements may seem to illustrate more than one fallacy.

1. It's so tragic to be blind; let us rescind the taxes of all blind people.
2. College is great. You get to meet new people, you get to live on your own; nobody bosses you anymore, you put off working for another four years. It's fun.
3. All college students are excellent scholars.
4. Down with taxes!
5. Of course students will enjoy the new course! It was designed for students to enjoy!
6. How can we trust you when you hang out with known delinquents?
7. Fenster may not win many points, but his one good point is hard and brilliant like a diamond.
8. You can't accuse me of plagiarism when you yourself are guilty of handing in work not your own.
9. Right after the Camp David Summit, the President's top Middle East advisor died; the strain must have killed him.
10. I don't know why you think Smith has sober judgment when you admit that he is a heavy drinker.
11. Either you memorize the fallacies or you fail the test; it's that simple.
12. I don't care what you say; I won't listen to a man who owns a porno theater.
13. If you are a good Republican, you will support the Republican platform.
14. "If they make my legs look this good, imagine what they will do for yours," said Joe Namath in his pantyhose ad.
15. We aren't going to commit our time and money to some drippy proposal from a place like Pretentious State University!

16. It's polite to speak quietly in the presence of the king; thus I shall be even more polite by speaking more quietly still; and perhaps I shall be the most polite of all by remaining silent!
17. Should innocent young children be made to suffer so that greedy capitalistic doctors can make exorbitant profits?
18. My Aunt Tillie refused to take communion wine for fear of becoming an alcoholic!
19. It is the old Christmas trees that bring the robins back to the North; no robin has ever been seen until after the Christmas trees have been tossed out.
20. No one has ever seen the inside of the earth, so it's at least possible that there may be people in there.
21. It is little wonder that so many students did poorly on the test; the professor gave us only a week to prepare.
22. Each spring, the government makes sure there are plenty of stories in the news about tax evaders and cheaters and what happens to them; the idea is to make people afraid of the IRS.
23. We might be more inclined to experiment with collective farming if this were not so blatantly a communistic approach to agriculture.
24. A surprising 90 percent of the young people in a recent study said they don't vote in elections. What has happened to America's youth?
25. My opponent wishes to know how I voted on the abortion bill. Well, there are many bills in Congress. A congressman's life is a busy life. Let me tell you just what I have done this past week.

STUDY GUIDE
Chapter Four Evaluating Evidence

1. What is the most important component of research? Why?
2. What is primary evidence?
3. What is secondary research?
4. What is the standard procedure for secondary research?
5. True or false: Inconclusive evidence is valid. Why?
6. Skills of the lab and skills of the courtroom: Which are primary and which are secondary?
7. How does the weight of evidence influence the researcher's conclusion?
8. How does the source of evidence influence its credibility?
9. What are some problems with surveys and questionnaires?
10. What is the difference between a reliable and an unreliable source?

11. Should researchers make use of encyclopedias?
12. Why shouldn't college students cite encyclopedias?
13. How might a source bias be built into a research question?
14. Why should researchers try to get a representative sampling of the kinds of sources available?
15. How might a researcher be led into the genetic fallacy?
16. True or false: In most serious research questions, the experts on both sides are likely to have good credentials.
17. What difference does it make who the experts are?
18. What is evidence?
19. What is testimony?
20. What is hearsay evidence?
21. What is an inference?
22. What is "mere" opinion?
23. What is critical judgment?
24. What is expert testimony?
25. What is *prima facie* evidence?
26. What is the meaning of *de facto*?
27. What is the meaning of *ex post facto*?
28. What is the *a priori* mistake in research?
29. What is the meaning of *ipso facto*?
30. What is the rule of definition?
31. True or false: Newer data are better than older.
32. Why might there be a break in the research between (for example) 1983 and 1988?
33. What is the traditional chronology of research?
34. What is Occam's Razor?
35. What is the rule of impartiality?
36. How might a research question have a built-in bias?
37. Should students use firsthand material in research papers?
38. What is common knowledge?
39. What is the rule of relevance?
40. What is the danger in using probable evidence?
41. What is statistical probability?
42. What problems arise with the use of statistical data?
43. What problems may arise from sampling techniques in statistical research?
44. Does Kluckhohn (#1) deny that apes can talk? What does his last sentence mean, in plain English?

45. What does Katz mean by "novel sentences"? If apes could produce "novel sentences," would Katz say they could "talk"?
46. Why is the Reynold's quote "not . . . at all relevant"?
47. In what way is Droscher's discussion of animal language behavior different from Kluckhohn's and Katz's? Why wouldn't the two earlier researchers agree with Droscher's meaning of communication?
48. In what ways are Sarah's and Washoe's use of language different from that in Droscher's excerpt? Would Kluckhohn or Katz agree that Sara and Washoe were using language? How strong is this evidence on a scale from conclusive to inconclusive? Why?
49. Why might there be "deep anxieties" summoned by the idea that a chimpanzee might be capable of language?
50. The last sentence in Linden (#8) explains the meaning of Lucy's sign "out baby." Is this explanation objective? Are there other examples of this kind of interpretive statement in Linden's work? What is the importance of the fact that Linden relies on Premack and the Gardners for some of his data?
51. What point is being made about word order in Baur (#9)? What does he mean by "semantic intentions"? How important is this objection?
52. Seidenberg and Petitto (#10) are the most recent researchers here; do they believe apes can talk? Are they right?
53. If these first ten items were all the research you had, what would you be willing to conclude? Can apes talk? (Why?)
54. What is the theoretical significance of this research question? Why is it important to know whether apes can talk?
55. Explain the differences among *assumption*, *inference*, and *premise*.
56. Reaching conclusions based on too little evidence is called what? Give an example.
57. What is card stacking? Give an example.
58. What is Occam's Razor and what is its relation to *argumentum ad ignorantium*? Give an example of an appeal to ignorance.
59. What is the meaning of *post hoc ergo propter hoc*?
60. Persuasion by force relies on what appeal?
61. Explain *ad hominem* and give an example.
62. What is the difference between *ad hominem* appeals and the genetic fallacy or fallacy of opposition?
63. Explain and give an example of guilt by association.
64. What is *ad misericordiam*? Give an example.
65. What is an *ad populum* appeal? What is most objectionable about populism?
66. Explain *ad verecundiam*; give an example.

67. What is bandwagon? What are its two subdivisions?
68. What is wrong with appeals to common sense?
69. An attack on a weak, minor, or irrevelent point is called what?
70. What is *tu quoque*? Give an example.
71. What is *amphiboly*? Give an example.
72. Explain begging the question. Why is it begging the question to cite God's word to prove the existence of God?
73. Explain equivocation and false metaphor; give examples.
74. What is a complex question or complex issue? Give an example.
75. What is a dilemma? What is a false dilemma?
76. Explain *non sequitur*; give an example.
77. What is a rationalization? Give an example.
78. Explain *reductio ad absurdum*; give an example.
79. Explain slippery slope. Why is it called a fallacy?

CHAPTER FIVE

Documentation

Documentation refers to the method by which writers tell their readers where information came from. Footnotes, endnotes, and in-text citations are various styles of documentation. Since other researchers must be able to evaluate the quality of your work, it's important to be accurate and consistent in the way you give citations. A citation note is a parenthetic name and/or page number that refers to an entry in the bibliography (called Works Cited or References) at the end of your paper. There are various documentation styles used by researchers today: MLA, APA, traditional footnotes, endnotes, and several others. If some problem should arise that isn't covered by the models in this chapter, you should make a reasonable adaptation from the most appropriate model. While documentation is essential in research writing, you must not let it overwhelm or distract from your text. The most efficient documentation is usually just to give attribution as part of your sentence: "In *Nim*, Terrace says chimpanzees do not take turns, do not act like they are participating in conversation."

HOW MUCH DOCUMENTATION?

For most research you should have as much documentation as possible. In some other types of writing, source material may be kept to a minimum. Students writing about literature, for example, are asked to rely on their own ideas and not quote extensively from critics and authorities. Then, too, well-known authorities often do not quote very much from other authorities. Nevertheless, for most research, you should have maximum documentation. In effect, your citation notes are the data. They point to the evidence in your research. It's normal for student research to be made up almost entirely of source material summarized, paraphrased, or occasionally quoted from authorities; you can rarely have too much documentation. A string of quotes makes for hard reading, of course (the string-of-pearls effect described below), but you can overcome this problem by paraphrasing more, summarizing the data in your own words, and working your evidence into a readable report. All such evidence would of course need citation notes.

RESEARCH PROBLEMS TO AVOID

The String of Pearls

Forcing the reader to read quote after quote, with little intervening explanation, is the string-of-pearls effect. Paraphrase and summarize in your own words to cut down on the number of direct quotes (but remember to document paraphrases and summaries). You cannot have too much documentation, but you can have too many quotes in your paper; use paraphrase instead. (See Substantiation, p. 8.)

Underresearched Paper

Make sure there are enough data. If you offer too little evidence, it will seem that you haven't done sufficient reading on the subject or that you have missed important books or articles. The general rule on how much is enough is to get all there is. Two dozen books and articles may seem like a lot to read, but if there are a hundred or more in the library, two dozen is less than 25 percent of the available material. You must make sure that your thesis is very thoroughly supported and that you have covered all the major aspects of your research question adequately.

Overworking the Data

Don't use a single example to illustrate two different points. That is, lyrics cited in one paragraph as an illustration of originality shouldn't be cited again in a later paragraph as an illustration of religious motif. Overworking the data suggests insufficient research. Find new examples for each new point you make, if possible.

Underdocumentation

If there aren't enough citation notes, the researcher is obviously quoting, paraphrasing, or plagiarizing without sufficient documentation. There shouldn't be undocumented information in the paper. It's better that every sentence have a citation note than to risk the appearance of "spotty" or casual documentation.

Plagiarism

Plagiarism is use of material without documentation. Plagiarism suggests intellectual dishonesty; practically speaking, it undermines research by confusing or obscuring the sources of information. There are two basic kinds of plagiarism: intentional and unintentional. The first is easy—and easy to spot. It involves the wholesale copying of sentences, paragraphs, even entire papers and calling them your own. Paraphrasing without acknowledging the source falls into the same category. Both are dangerous, dishonest practices.

Unintentional plagiarism is different; it could arise from failure to put quotes around material in your notes or from a lack of understanding of what should be documented. Be very careful in this area because your instructor has no way of knowing the cause of plagiarism in a paper. Give a parenthetic citation for anything in your paper that your readers may believe came from a book or article, especially every indirect quote, paraphrase, unconscious borrowing. (See "What To Document," pp. 148–150.)

REFERENCE NOTES

Once the most popular style of documentation, footnotes and endnotes are less widely used today (see the end of this chapter, p. 196). In place of the traditional reference notes, many disciplines now use one of the in-text citation styles. A parenthetic citation is given in your paper, like this (Smith 97). The number (97) is a page reference, and your reader will find Smith's work listed in the bibliography at the end of your paper.

CONTENT NOTES

A content note is a note from you to your readers. (See page 339 for an example.) The note gives additional information or clarification. There is some bias against using such notes; many researchers feel that any relevant or important information ought to be included in the body of the paper, and anything else should be omitted. However, since researchers do sometimes use substantive (content) notes, the best advice is to use such notes only when they would otherwise seem intrusive in your paper. Avoid showing off with content notes that supply all sorts of incidental information.

In the in-text documentation style, content notes (and only content notes) are treated as endnotes. Give a note number at the appropriate place in your paper, raised half a line. At the end of your paper, ahead of your Works Cited or References page, list your content notes, with corresponding numbers, on a Notes page. See p. 339 for a model Notes page. Some researchers feel that extensive notes indicate the thoroughness of their scholarship, but most student work should have few content notes. If you have more than three, you should reconsider what you are doing: too much material is being treated as incidental when its sheer quantity seems to suggest otherwise. Possible content notes include the following:

Identification You provide relevant information to identify a person, place, or object mentioned in your paper.
Explanation You clarify a statement, especially as quoted from a source, by adding more information.
Definition You define a term or add to the definition in your paper.
Extension You refer the reader to other works, other sources, for further information on a point.
Evaluation You comment on the relative quality of a source or compare one with another.
Cross-reference You direct readers to see another part of your paper for additional information.
Opinion You offer your own opinion, doubt, or denial where a reader might mistakenly assume you agree with ideas or words in source material.

WHAT TO DOCUMENT

Direct Quotations

Any words you copy from a source should be placed in quotation marks, followed by a parenthetical citation, followed by the sentence period.

> The results show "a measurable decrease in oxygen" (Leng 47).

Words and Ideas from a Source

A word or two or a concept that you take from a source and incorporate into your own sentences should be documented:

> Never at a loss for words, Holden manufactures adjectives from nouns and verbs to describe "vomity" taxicabs and "hoodlumy-looking" (98) street people.

Changing the words doesn't change your obligation to document. Ideas, interpretations, analyses, and concepts that might seem to be conclusions on your part should be documented. For most research it's both acceptable and desirable for you to document everything in your paper. Your research is valid if your own conclusions merely confirm the conclusions of other researchers, but you must avoid giving the impression that you are claiming for your own, ideas originated by others. Other researchers may believe you are stealing (borrowing without acknowledging the source) ideas to make yourself seem authoritative. Using information or ideas without documenting them is the definition of plagiarism. No researcher can afford the damage plagiarism does to credibility. You can document someone else's thinking with a content note:

> Holden can be thought of as a model of lost and confused adolescence,[1] but he is a prototype for a very small, privileged class of modern people.

On a Notes page, you would provide a note like this:

> [1] See Sara Birkfeld, ''A Jungian Look at Catcher in the Rye,'' Psychology Today, May 1975: 72–77, for the interpretation of Holden Caulfield as an archetype.

Paraphrases and Restatements

When you change words, phrases, or ideas into your own language, you must still give a note:

> Washoe was said to be able to use language to describe conditions other than reality; she could tell a lie. When asked who had soiled the carpet, Washoe indicated her trainer had done it (Linden).

Allusions and Incomplete References to Sources

Regardless of whether you quote from them or paraphrase them, document all allusions and incomplete references:

> Broder's Changing of the Guard gives a disturbing picture of the changing ideas, values, and leadership in America.

No additional information is needed in your paper, but in your Works Cited list, you must have the following:

> Broder, David. Changing of the Guard. New York: Simon, 1980.

Major Source

If you make many references to the same source—for example, if you are writing about a book, poem, play, or other source—you needn't keep documenting it. As long as it's clear to the reader that you are still talking about the same source, you may use only page numbers for references.

In Silent Partners, Linden says that chimps look like they are trying to communicate (30), but ''whether she [Washoe] was using language is another question'' (31). Is communicating the same as using language? Linden suggests it would be difficult to prove that a human infant really has language:

> At eighteen months [his daughter] began to demonstrate most of the behaviors critics have cited as evidence that apes cannot learn language: she interrupted, she repeated phrases incessantly, she responded inappropriately, and she jumbled her word order. (34)

But the human child steadily and rapidly progresses beyond this stage (35). Chimps do not.

Source within a Source

There is a very strong prejudice against using material you find quoted in a secondary source; the general rule is simply don't quote a source in a source. Always find the original. For example, if you find Smith quoting Jones, you must not use Smith's quote unless there is some very good reason why you can't find the original source by Jones. If the original source is out of print or simply impossible for you to find, then and only then is it acceptable to quote from the secondary source:

MLA
Washoe's trainers were ''whispering into a recorder'' (qtd in Ristau and Robbins 164).

APA
Washoe's trainers were ''whispering into a recorder'' (cited in Ristau and Robbins, 1982, p. 164).

WHAT NOT TO DOCUMENT

Common Knowledge

Common knowledge needn't be documented. Knowledge is "common" when it's widely known by educated people. Within the specialized area of your research, knowledge is "common" if it's readily available in most general reference works such as encyclopedias and almanacs or through the popular media—television, newspapers, and magazines. (See the section on Common Knowledge in Chapter 4.) There is no need to document, for example, who the president is or where the White House is or the fact that Shakespeare is the author of *Hamlet*. (Anything you *copy* from a source you must of course document, even if it is common knowledge.)

Uncontested knowledge needn't be documented, even if it isn't common knowledge. Dates of historical events, for example, may or may not be considered common knowledge. But unless the precise date is relevant, or unless the dates are a matter of dispute in the research, they can be considered uncontested information. A handy rule to follow is this: anything that would damage your case if it were removed from your paper or proved to be wrong should be documented.

When in doubt, document. All data used by you to establish a thesis should be documented. Since students usually aren't authorities, you must avoid using undocumented data to appear to know more than you do.

ACTIVITY 28 Read the following excerpt and the statements that follow it. Using the researcher's rule—if you found it, cite it—indicate whether you think researchers would require citation notes for the statements. Explain why the statements do or do not need citations, in your opinion.

1

COCKROACH

Cockroach, the name applied to members of the Blattidae, a family (sometimes considered an order, Blattaria) of orthopterous insects, with flattened bodies, long, threadlike antennae and shining leathery integument. The name is a corruption of the Spanish *cucaracha*; in the U.S. it is commonly shortened to ''roach.''

They are chiefly tropical, but certain species have become widely disseminated through commerce and are now cosmopolitan. Cockroaches are nocturnal in habit, hiding themselves during the day. The domestic species are omnivorous but are especially addicted to starchy or sweetened matter of various kinds; they also attack food, paper, clothing, books, shoes, bones, etc., and dead insects. As a rule cockroaches damage and soil far more than they consume, and many species emit a disagreeable odour.

The oriental cockroach (*Blatta orientalis*), a cosmopolitan household pest, is dark brown; males are short-winged, females vestigial-winged. The larger, fully winged American cockroach (*Periplaneta americana*) infests buildings throughout the tropics and warm-temperate zones. The German cockroach (*Blatella germanica*), small and pale with two dark lines on the pronotum, occurs with man from the tropics to high latitudes. Eggs of most cockroaches are laid in cases (oöthecae) which are carried protruding from the female's body until hidden in some crevice. Oöthecae of *P. americana* contain 10 to 16 eggs, which hatch in 40–45 days; nymphal life is 11–14 months, adult life 3–12 months. *B. orientalis* lives about one year.

Of the about 1,600 species of cockroaches known, only 62 species, many introduced, occur in North America; Great Britain has only 2 native species. Although blattids are usually drably coloured, some tropical species are elegant in form and beautiful in coloration. Some species are giants with a wingspread of more than five inches.

Although these insects are usually viewed with disgust, they are not devoid of interest. They are the most primitive of living winged insects, and are among the oldest fossil insects. Their generalized structure and large size make them convenient for study and dissection, and they are widely used as the most suitable type for commencing the scientific study of insect morphology.

Insecticides used for cockroach control, most of which are poisonous to animals and man, include chlordane, one of the most effective; DDT; pyrethrum; and sodium fluoride, which was used for many years.

Theodore H. Hubbell, ''Cockroach,'' Encyclopaedia Britannica, 14th edition,

1. Hubbell says: "As a rule cockroaches damage and soil far more than they consume, and many species emit a disagreeable odour."
2. There are no winged insects more primitive than cockroaches.
3. The cockroach doesn't lay eggs the way a chicken does, but instead lays a little pouch or case containing as many as a dozen eggs.
4. Roach eggs hatch in approximately a month and a half, and some roaches live as long as a year.
5. Cockroaches are of course insects, similar to many creatures we call "bugs."
6. Not all roaches are drab and ugly; some found in the tropics are quite elegant and beautifully colored.
7. The scientific name for our common roach is *Periplaneta americana.*
8. The ordinary roach looks a good deal like other insects, such as crickets, grasshoppers, and certain beetles.
9. The domestic cockroach will eat nearly anything, including nonfood items.
10. The cockroach is the subject of the well-known Spanish song, "La Cucaracha."
11. There are a lot of poisonous insecticides that have been used against the cockroach, many of which will harm animals and people as well as cockroaches. These substances include sodium fluoride, pyrethrum, and DDT, but the most effective poison is chlordane.
12. Since there are so many cockroaches everywhere, we will never be able to get rid of them completely.
13. Cockroaches are described as creatures that are basically "nocturnal"; this word means they hide themselves during the day.
14. Students who are beginning to study the physical structure of insects often find that the cockroach is the best insect to start with.
15. Probably the world's best known cockroach was Archy of *Archy and Mehitabel* by Don Marquis.

Do Not Document Paragraphs

Don't document a whole paragraph of paraphrased material. If you wish to quote an entire paragraph, you may do so with an indented quote. (See the Punctuation section in Chapter Seven.) Another possibility is to acknowledge your source at the beginning of the paragraph. For example: "Smith gives the relevant statistics in the following paragraph (my paraphrase)." Otherwise, don't attempt to "change the words a little" for whole paragraphs. Avoid writing any paragraph so that the last sentence ends with a parenthetic citation. The reader will be unable to tell whether the citation refers to the sentence or to the entire paragraph. If you must document the last sentence in one of your paragraphs, make sure that the citation falls within the sentence, not at the end, outside the end punctuation. It is better that every sentence have a citation than that large blocks of information look undocumented except for a single note at the end.

IN-TEXT RULES

The in-text documentation style (MLA) replaces footnotes or endnotes with citations inside your paper itself. Enough information is given in your text so that a reader can find the source in the Works Cited list. Thus, much documentation can be reduced to an author's name:

> Seidenberg and Petitto note the conflict in the scientific community over whether ape behavior constitutes language in a human sense.

Direct quotations can be identified with a parenthetic reference to name and page number:

> ''These apes likewise signal 'Come here' or 'walk quickly past me,' with gestures that are amazingly human'' (Droscher 208).

This in-text style does away with notes (except for content notes to the reader), and it's meant to keep documentation conveniently close but unobtrusive. Learn how to incorporate documentation smoothly into your writing just as you learn to incorporate the words and ideas of your sources. The next few sections give some guidelines.

Use Author's Name

Most of your references should give the author's name in your text instead of in a note.

> According to Ashley Montagu, the gorilla Koko can create metaphors, swear, lie, express contrition and other emotions, and, to some extent, understand the English spoken to her.

In the name and date style, include the publication date as well: According to Ashley Montagu (1982). . . . If you are using the MLA style, give all authors' names unless there are more than three. For the APA style give all authors' names (up to five authors) in the first reference, but only the first author's name and *et al.* thereafter.

A significant test of language—can chimps use signs spontaneously with <u>each other</u>?—has been passed, according to Fouts, Fouts, and Schoenfeld in ''Sign Language Conversational Interaction between Chimpanzees.''

Apes have passed a significant language test: they have begun to use language to communicate with each other (Fouts, Fouts, and Schoenfeld).

For more than three authors, give only the first name followed by *et al.* in MLA style; give only the first author and *et al.* for six or more authors in APA style.

How the chimps are housed and raised—in cages or free to move about—significantly influences their language learning (Mathieu et al.).

For authors with the same last name, give full names in your text, or give the authors' initials in a citation note (unless the initials are the same).

David Premack poses this question: Why is there no language link between man and his nearest relative, the chimpanzee?

In <u>Why Chimps Can Read</u>, Ann Premack says if we invented a simple language, in which a label was a ''word'' and words consisted of names of chimps, people, actions, places, food, we could place these labels on a table, ''write'' to the chimpanzee, and the chimpanzee could ''read'' and reply.

Use Name and Title

If there is more than one work by the same author in your bibliography, you must include the titles in your references.

In ''In the Beginning was the 'Name,''' Terrace says the ability to ''refer''—the ability to name (without an intention to obtain)—is at least as important as other language skills, in fact, more important than other skills. And in <u>Nim</u> he says Nim was ''observed to gaze at his fingertips while making a sign in the absence of the sign's referent'' (143).

Use Page Numbers

When citing a specific part of a source, you must give a page number in the citation. When the page number comes at the end of your sentence, the sentence period should be placed after the parenthesis. Note that in MLA style, page numbers aren't identified with *p.* or other markers. If additional references are made to the same source, you need only the page number. Don't precede page numbers with a comma. However, see APA guidelines for treatment of page references in the name and date style.

MLA

The child imitates the adult's use of language, almost immediately produces sounds and words without cues from adults (Savage-Rumbaugh 25). Apes almost never do, must be cued, coaxed heavily. Chimps use eating language to request food. The child can use it to comment on attributes of food, eating behaviors, actions of others (mother pouring, e.g.): ''This is a dramatic and significant difference between use of symbols in children and in chimpanzees, and is obvious even at the earliest stages of acquisition'' (25-26).

APA

The child imitates the adult's use of language, almost immediately produces sounds and words without cues from adults (Savage-Rumbaugh, 1986, p. 25). Apes almost never do, must be cued, coaxed heavily. Chimps use eating language to request food. The child can use it to comment on attributes of food, eating behaviors, actions of others (mother pouring, e.g.): ''This is a dramatic and significant difference between use of symbols in children and in chimpanzees, and is obvious even at the earliest stages of acquisition'' (pp. 25-26).

Inclusive Page Numbers and Dates

To show inclusive page numbers, MLA style uses full numbers from 1 to 99: 8–13; 29–33; 95–99. For numbers larger than 99, shorten the second number: 103–06; 235–41 (but avoid confusion: 197–205; 399–401). Use the same principle with inclusive dates: 1988–89; 1899–1901.

APA style does not shorten numbers or dates: 103–106; 235–241; 1988–1989.

Use Shortened Titles

To keep references as brief as possible, shorten titles, but make them unambiguous so the reader can recognize the titles in your Works Cited. For example, references to Terrace's work "In the Beginning was the 'Name,'" could be shortened:

Terrace argues that in most of the research, apes ''use'' language to request things, they do not use it to ''refer''—to name without requesting (''In the Beginning'').

For references to books of the Bible, use standard abbreviations. (See Chapter 6.)

> Seek ye out of the book of the Lord, and read. . . . (Isa. 34.16).

For references to plays, poetry, or other works with numbered sections or lines, give all the relevant numbers that would help a reader find the source: section, part, act, scene, line. Don't use *l.* or *ll.* for *line.*

> O, what a rogue and peasant slave am I! (Hamlet II.ii.534).

The line is from *Hamlet*, Act two, Scene two, line 534. You may use arabic numerals if you prefer: 2.2.534.

> An aged man is but a paltry thing, / A tattered coat upon a stick . . . (''Byzantium'' 9–10).

The lines are from Yeats's "Sailing to Byzantium," lines 9–10.

BIBLIOGRAPHY

Always take bibliographic information from the title page and copyright page of your source, not from indexes or the card catalog.

Authors' Names

Take authors' names from the title page. Don't abbreviate or substitute initials for full names. In MLA style, give the name exactly as it appears on the title page. (See APA Guidelines for a different rule.)

Titles

Give exactly as given on title page. Include subtitles. Disregard articles *a*, *an*, and *the* when alphabetizing authorless titles:

> ''The Chimps Have the Last Word,'' Midland Times, 16 Apr. 1975, Sec. 3:5.
>
> ''The Planet of the Apes?'' Rochester Clarion, 28 Oct. 1977, Sec. B:3.

Place of Publication

Use only the first city if several are listed. If there is likely to be some ambiguity (London, Ontario), give state or province as well. Use postal abbreviations for state names.

Shorten Publishers' Names

Writers and others in the publishing world quickly become familiar with the major publishers' names; it's usually not necessary to give the full name. The general rule is to give that part of the name that people would use to look up the publisher's address. Thus, words and abbreviations like *Company, Co., Inc.* and so forth can be dropped. Names formed from two names (Harcourt Brace, Houghton Mifflin, Prentice-Hall, etc.) need only the first name: Harcourt, Houghton, Prentice, and so on.

Professional organizations are frequently referred to by their initials (NCTE, MLA, etc.), and you may list them that way in your Works Cited list. But if the organization is likely to be unfamiliar to your readers, spell out the name in full (National Council of Teachers of English, Modern Language Association).

Some publishers have divisions with different names. Anchor is a division (imprint) of Doubleday. Use a hyphen to separate division names from parent company: Anchor-Doubleday.

Copyright Date

The copyright date is frequently listed on the back of the title page. The copyright date is usually the date of the first edition, but it isn't necessary for you to include the words "First Edition" along with the date. Only editions other than the first need to be identified.

BASIC BIBLIOGRAPHY FORM, BOOK (MLA)

This basic bibliography form for books is standard for the new MLA as well as the traditional documentation style.

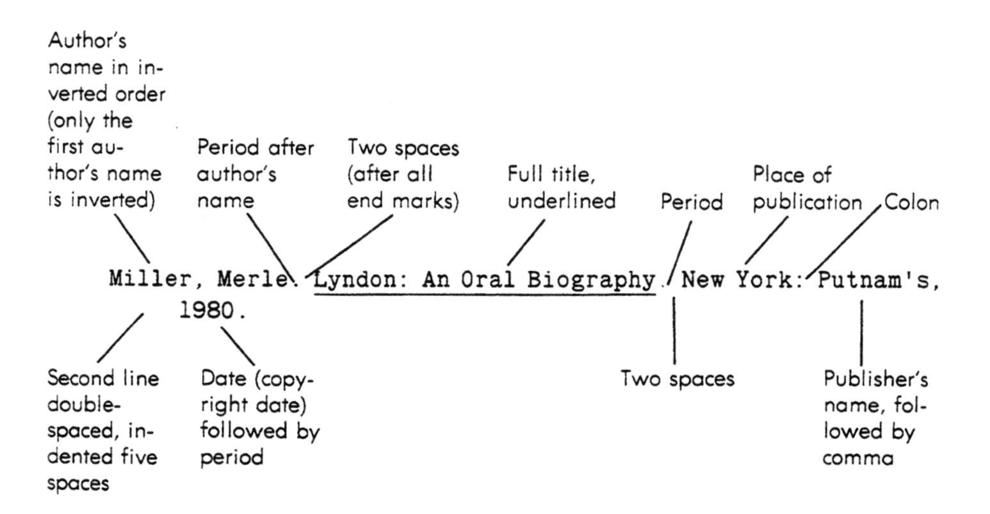

BASIC BIBLIOGRAPHY FORM, PERIODICAL (MLA)

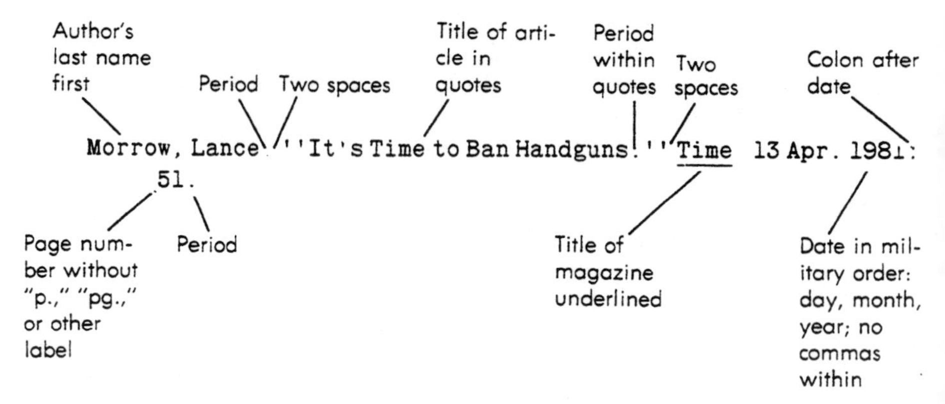

MODEL BIBLIOGRAPHY (WORKS CITED) FORMS: BOOKS

ONE AUTHOR

Allan, Charles M. The Theory of Taxation. Baltimore: Penguin, 1971.

Mailer, Norman. The Executioner's Song. Boston: Little, 1979.

Pressman, Roger S. Numerical Control and Computer-Aided Manufacturing. New York: Wiley, 1977.

Wolfe, Tom. The Right Stuff. New York: Bantam, 1980.

Note that items in a bibliography are in alphabetical, not numerical, order.

MORE THAN ONE BOOK BY SAME AUTHOR

Don't repeat an author's name with subsequent books in MLA style. When there are two or more books by the same author, a string of three hyphens takes the place of the author's name. The hyphens indicate *exactly* the same authorship as the listing above. Any different information about the authorship of subsequent books must be indicated. If the author is an *editor, translator, or served in any capacity different from that in the first book* of a subsequent book, this new information must be included (---, ed.)

Note that this convention doesn't apply to coauthorships. If Postman has written one book and coauthored another, don't use hyphens for Postman's name with the coauthored book.

Kennedy, George A. The Art of Persuasion in Greece. Princeton, NJ: Princeton UP, 1963.

——. The Art of Rhetoric in the Roman World, 300 B.C.–A.D. 300. Princeton, NJ: Princeton UP, 1972.

——. Classical Rhetoric and Its Christian and Secular Tradition from Ancient to Modern Times. Chapel Hill: U of North Carolina P, 1980.

AUTHOR OF ONE BOOK, COAUTHOR OF ANOTHER

Postman, Neil. Teaching as a Conserving Activity. New York: Delacorte, 1979.

Postman, Neil, and Charles Weingartner. Teaching as a Subversive Activity. New York: Delacorte, 1969.

TWO OR MORE AUTHORS

Note that only the first author's name is inverted. The names of second and third authors are given in normal order.

Downie, N. M., and R. W. Heath. Basic Statistical Methods. 2nd ed. New York: Harper, 1965.

Field, Frank, Molly Meacher, and Chris Pond. To Him Who Hath: A Study of Poverty and Taxation. New York: Penguin, 1977.

If there are more than three authors, you may use only the first author's name with "et al." ("and others"). If you prefer, you may instead give all authors' names, and you must give all the names in your bibliography if you refer to the others by name in your text. For example, the title page of *Murmurs of Earth* (Random House, 1978) lists six authors: Carl Sagan, F. D. Drake, Ann Druyan, Timothy Ferris, Jon Lomberg, and Linda Salzman Sagan. In your Works Cited list, give the first author's name, followed by "et al.":

Sagan, Carl, et al. Murmurs of Earth: The Voyager Interstellar Record. New York: Random, 1978.

Sagan, Carl, F. D. Drake, Ann Druyan, Timothy Ferris, Jon Lomberg, and Linda Salzman Sagan. Murmurs of the Earth: The Voyager Interstellar Record. New York: Random, 1978.

MULTIPLE PUBLISHERS

Phelps, Richard H. Mass Violence in America: A History of Newgate Connecticut. New York: Arno Press & The New York Times, 1969.

New Technology for Libraries: A Layman's Guide to Reducing Public Library Costs and Improving Services Through Scientific Methods and Tools. A Background Paper for the White House Conference on Library and Information Services. New York: National Citizens Emergency Committee to Save Our Public Libraries; Washington: National Commission on Libraries and Information Science; American Library Association; Urbana Libraries Council, 1979.

COMMITTEE OR GROUP AUTHOR

Committee on the National Interest. The National Interest and the Continuing Education of Teachers of English. Champaign, IL: NCTE, 1964.

If you have several books by the same group author, you may use an alternative listing:

The National Interest and the Continuing Education of Teachers of English. By the Committee on the National Interest. Champaign, IL: NCTE, 1964.

COMMITTEE OR GROUP AUTHOR NAMED IN TITLE

It's desirable to show an author's name whenever possible, even though the name may be repeated in the title.

The Warren Commission. Report of the Warren Commission on the Assassination of President Kennedy. New York Times Edition. New York: McGraw, 1964.

COMMITTEE OR GROUP, AUTHOR SAME AS PUBLISHER

Canadian Tax foundation. Provincial and Municipal Finances. Toronto: Canadian Tax Foundation, 1979.

BOOK WITH EDITOR(S)

Hartman, Frederick H., ed. World in Crisis: Readings in International Relations, 4th ed. New York: Macmillan, 1973.

Bullock, Allan, and Oliver Stallybras, eds. The Harper Dictionary of Modern Thought. New York: Harper, 1977.

Note no additional period is needed after abbreviations like "eds."

BOOK WITH EDITOR AND AUTHOR

Twain, Mark. Letters from the Earth. Ed. Bernard Devoto. New York: Harper, 1962.

CHAPTER OR SECTION OF BOOK BY ONE AUTHOR

Asimov, Isaac. ''What Is Science?'' Asimov's Guide to Science. New York: Basic, 1972.

ESSAY, CHAPTER, OR SECTION IN AN EDITED WORK

Kline, Morris. ''The Meaning of Mathematics.'' In Adventures of the Mind, 2nd Series. Eds. Richard Thruelsen and John Kobler. New York: Vintage, 1972. 77–91.

Note use of page numbers here.

TRANSLATION, AUTHOR'S NAME FIRST

Include the translator's name if it appears on the title page.

Dostoevsky, Fyodor. Crime and Punishment. Trans. Constance Garnett. New York: Modern Library, 1950.

Put the author's name first if you are writing about the book itself.

TRANSLATION, TRANSLATOR'S NAME FIRST

Garnett, Constance, trans. Crime and Punishment. By Fyodor Dostoevsky. New York: Modern Library, 1950.

The translator's name comes first if you are writing about the translator's work, such as the accuracy of the translation.

MULTI-VOLUME WORK

Churchill, Winston S. The Second World War. 6 vols. Boston: Houghton, 1948–53.

PART(S) OF MULTI-VOLUME WORK

Churchill, Winston S. Triumph and Tragedy. Boston: Houghton, 1953, Vol. 6 of The Second World War. 6 vols. 1948–53.

Steffens, Lincoln. *The Autobiography of Lincoln Steffens*. Vol. II. New York: Harcourt, 1958. 2 vols.

Note that Churchill's work has individual titles and publication dates for each volume; Steffens's autobiography has one title for both volumes.

In-text references to part of a multi-volume work must give volume number and page number (6: 101–02).

ANONYMOUS (AUTHOR UNKNOWN)

Don't use "Anonymous" or "Anon." when authors' names are unknown.

Beowulf. Trans. William Ellery Leonard. Illus. Lynd Ward. Norwalk, CT: Heritage, 1967.

Note that the illustrator is often important to the success of a book, especially, as in this edition of *Beowulf*, when illustrations have been added to material not illustrated in the original. Give illustrators' names when listed on title page.

The Book of Kells: Reproductions from the Manuscript in Trinity College Dublin. New York: Knopf, 1974.

Always give complete titles in your bibliography. Note the confusion that might arise without the subtitle to this edition of *The Book of Kells*.

ANONYMOUS, NAME SUPPLIED

[Rudolph Raspe et al.]. *The Singular Adventures of Baron Munchausen*. Norwalk, CT: Heritage, 1952.

Always use the name listed on the title page, including the supplied name of an anonymous work. If no name is listed, begin your entry with the title. Don't supply names not given on the title page unless it is important for the reader to have that information or you refer to the names in your paper.

PSEUDONYM (PEN NAME), NAME SUPPLIED

Mark Twain [Samuel Langhorne Clemens]. *The Autobiography of Mark Twain*. Ed. Charles Neider. New York: Harper, 1959.

SERIES

Brée, Germaine, ed. *Camus, A Collection of Critical Essays*. Twentieth Century Views. Englewood Cliffs: Prentice, 1962.

Loban, Walter. *Language Development: Kindergarten Through Grade Twelve*. NCTE Research Report 18. Urbana, IL: NCTE, 1976.

Peyre, Henri. ''Camus the Pagan.'' *Camus, A Collection of Critical Essays*. Twentieth Century Views. Ed. Germaine Brée. Englewood Cliffs: Prentice, 1962. 65–70.

Thorp, Willard. *American Humorists*. University of Minnesota Pamphlets on American Writers 42. Minneapolis: U of Minnesota P, 1964.

Note that "University" and "Press" as parts of publishers' names are given as "U" and "P" in MLA style.

MONOGRAPH

Bryan, R. D. *Reactions to Teachers by Students, Parents, and Administrators*. Kalamazoo: Western Michigan U, 1963.

Coleman, J. S. *Social Climates in High School*. Cooperative Research Monographs 4. Washington, DC: U.S. Office of Education, 1961.

Zimmer, K. E. *Affixal Negation in English and Other Languages*. Monograph No. 5, Supplement to *Word* 20, 1964.

REPRINT OF OLDER WORK

Agee, James, and Walker Evans. *Let Us Now Praise Famous Men*. 1939. New York: Ballantine, 1966.

EDITION

It isn't necessary to identify first editions. But for all subsequent editions, give an edition number in your Works Cited list.

Gunther, Erna. *Ethnobotany of Western Washington: The Knowledge and Use of Indigenous Plants by Native Americans*. Rev. ed. Seattle: U of Washington P, 1973.

Leggett, Glenn C., David Mead, and William Charvat. *Prentice-Hall Handbook for Writers*. 8th ed. Englewood Cliffs: Prentice, 1982.

INTRODUCTION, PREFACE, FOREWORD

Cowley, Malcolm, ed. ''How Writers Write.'' Introd. *Writers at Work: The Paris Review Interviews*. New York: Viking, 1959.

Ernst, Morris L. Foreword. Ulysses. By James Joyce. New York: Random, 1934.

BIBLE

The Holy Bible. Authorized King James Version. New York: Harper, n.d.

The New English Bible with the Apocrypha. Oxford UP; Cambridge UP, 1970.

Don't underline Bible titles. Don't list the Bible in your Works Cited list merely to cite (in text) a biblical passage. A listing for the Bible is required only when discussing the Bible itself (the Bible as literature, for example) or when citing passages in other than the authorized version.

DICTIONARY

Webster's New Collegiate Dictionary. 8th ed. Springfield, MA: Merriam, 1973.

Webster's New International Dictionary of the English Language. 3rd ed. Unabridged. Springfield, MA: Merriam, 1961.

Don't cite the dictionary merely to define a word; give a reference in the text instead. However, in a paper discussing the dictionary itself, a complete bibliographic reference is appropriate.

ENCYCLOPEDIA, ALPHABETICALLY ARRANGED WORK

Page numbers are unnecessary for books with an alphabetical arrangement.

''Alger, Horatio.'' DAB. 1872

[*DAB* is *Dictionary of American Biography.*]

Archibald, Kenneth. ''Cerebral Palsy.'' Encyclopedia Americana. 1976 ed.

Chapell, Duncan. ''Crime and Law Enforcement.'' Britannica Book of the Year. 1980.

''Chickamauga Dam.'' Encyclopedia Americana. 1976 ed.

''Roosevelt, Anna Eleanor.'' American Women, Vol. III (1939–40). Rpt. 1974.

Thomson, William E. ''Sound, Musical.'' Encyclopaedia Britannica: Macropaedia. 1974 ed.

Britannica articles are "signed" with authors' initials. Find the authors' names in *Propaedia: Guide to Britannica.*

''Walters, Barbara.'' Who's Who of American Women. 11th ed. 1979-80.

DISSERTATION, UNPUBLISHED

Kitzhaber, Albert Raymond. ''Rhetoric in American Colleges 1850-1900.'' Diss. U of Washington, 1953.

DISSERTATION, PUBLISHED

Selby, Stuart Allen. The Study of Film as an Art Form in Secondary Schools. Diss. Columbia U, 1963. New York: Arno, 1978.

ARTICLE FROM DISSERTATION ABSTRACTS, OR DISSERTATIONS ABSTRACTS INTERNATIONAL

Hazen, C. L. ''The Relative Effectiveness of Two Methodologies in the Development of Composition Skills in College Freshman English.'' DAI 33 (1973): 4243A. North Texas State.

BIBLIOGRAPHY FORMS: PERIODICALS

WEEKLY MAGAZINE ARTICLE

Raskin, A. H. ''A Reporter at Large: After Meany.'' The New Yorker 25 Aug. 1980: 36-76.

Abbreviate months except May, June, July. Note that the day precedes the month, and no comma is needed to separate the date from the title or the month from the year. Do not use volume or issue numbers with magazines.

MAGAZINE ARTICLE, NO AUTHOR GIVEN

''Byrd of West Virginia: Fiddler in the Senate.'' Time 23 Jan. 1978: 12-16.

MONTHLY MAGAZINE ARTICLE

Peirce, Roger M. ''Your Dollar Is at a Premium.'' The Nation's Business Dec. 1980: 92-100.

Note that page numbers aren't identified by "page" or "p." or any other designator in this style.

NEWSPAPER ARTICLE

Talbert, Bob. ''Why Are We in Such a Downer?'' Detroit Free Press 10 Jan. 1978, sec. A: 9.

Include the edition number, if any, after the date ("early ed.," "city ed.," etc.).

NEWSPAPER ARTICLE, UNSIGNED

''Latest Bid from Iran Won't Free Hostages.'' Detroit Free Press 28 Dec. 1980, sec. A: 1+.

Note that the plus sign (+) is used to indicate that the article appears on discontinuous pages. Part of it is on page 1, but the rest of it appears on a later page.

REVIEWS, SIGNED AND UNSIGNED

Aire, Sally. Rev. of Othello. Royal Shakespeare Company. Stratford. Plays and Players 27.2 (1979): 23-24.

Gross, Cheryl W. ''Guilt and Innocence in Marathon Man.'' Rev. of Marathon Man. Literature/Film Quarterly 8.1 (1980): 52-68.

Note volume and issue numbers (8.1), necessary when each issue begins on page 1. See "Professional, Technical, or Specialty Journal" below.

Burgess, Anthony. Rev. of Running Dog, by Don DeLillo. Saturday Review 16 Sept. 1978: 38.

Maddocks, Melvin. ''Joan of Arc.'' Rev. of Blood Red, Sister Rose, by Thomas Keneally. Time, 10 Feb. 1975: 76.

Rev. of Let Us Now Praise Famous Men, by James Agee and Walker Evans. New York Times Book Review 4 Jan. 1981: 23.

Note that some reviews ("Joan of Arc") are titled.

EDITORIAL

''National Security, Absurdly Defined.'' Editorial. New York Times 1 June 1980, sec. 4: 20.

''Mr. Burger's Case for Prison Reform.'' Editorial. The Christian Science Monitor 30 Dec. 1980: 24.

The *Monitor* doesn't have sections.

LETTER TO THE EDITOR

Dillon, J. T. ''A Solution to the Uncertainties of Scholarly Publication: In Model Responses to West.'' Letter. College English 42 (1980): 405-08.

Sander, Alois W. ''Milliken Budget Cuts Don't Favor Bureaucrats.'' Letter. Detroit Free Press 31 Dec. 1980, sec. A: 6.

Taylor, John C. ''China Revisited.'' Letter. Life Dec. 1980: 44.

Walters, Barbara. ''Hotel Oloffson.'' Letter. People, Special Double Issue, 29 Dec./5 Jan. 1981: 6.

PROFESSIONAL, TECHNICAL, OR SPECIALTY JOURNAL, EACH ISSUE STARTING WITH PAGE 1

Clark, Blair. ''Notes on a No-Win Campaign.'' Columbia Journalism Review 19.3 (1980): 36-40.

Note volume and issue numbers (19.3). See APA guidelines for a different treatment of volume and issue numbers.

Paterson, Katherine. ''Creativity Limited.'' The Writer 93.12 (1980): 11-14.

''SFPA Hears Grim Forecast for World Economy, Hope for U.S.'' Forest Industries 107.13 (1980): 11-13.

Stephenson, Susie. ''Feeding the Navy's Crew.'' Food Management 15.12 (1980): 40-43.

PROFESSIONAL, TECHNICAL, OR SPECIALTY JOURNAL, PAGES NUMBERED CONTINUOUSLY THROUGHOUT VOLUME

Blum, Lenore, and Steven Givant. ''Increasing the Participation of Women in Fields That Use Mathematics.'' The American Mathematics Monthly 87 (1981): 6-22.

Furukawa, Masao. ''Wall Temperatures and Solar Deflection of Cylindrical Structures in Space.'' Journal of Spacecraft and Rockets 17 (1980): 501–14.

Garver, Eugene. ''Demystifying Classical Rhetoric.'' Rhetoric Society Quarterly 10 (1980): 75–81.

McDowell, Bart. ''The Aztecs.'' National Geographic 158 (1980): 704–51.

TITLES AND QUOTES WITHIN TITLES

Berthoff, Ann E. ''Response to The Students' Right to Their Own Language, CCC, 25 (Special Fall Issue, 1974).'' College Composition and Communication 26 (1975): 216–217.

Sweet, Charles A., Jr., and Harold R. Blythe, Jr. '''Try It: You'll Like It': A Primer for Educational Television in the Classroom.'' College English 39 (1978): 608–15.

GOVERNMENT PUBLICATIONS AND OTHER DOCUMENTS

CONGRESS

Cong. Rec. 19 June 1979: S7929–40.

Give the name of the government followed by the agency when the author of the document is unknown.

United States. Cong. House. Agreement with Univ. of Texas for Lyndon B. Johnson Presidential Archival Depository. 89th Cong., 1st sess. H. Rept. 892. Washington: GPO, 1965.

For additional documents by the same government and agency (United States Congress), use three hyphens in place of the names.

---. ---. House. Committee on House Administration. National Publication Act of 1980. 96th Cong., 2nd sess. H. Rept. 836. Washington: GPO, 1980.

---. ---. Conference Committee. 1980. Providing for an Extension of Directed Service on the Rock Island Railroad. 96th Cong., 2nd sess. H. Rept. 1041 to accompany S. 2253. Washington: GPO, 1980.

———. ———. Joint Committee on Taxation. Explanation of Proposed Estate and Gift Tax Treaty Between the United States and the United Kingdom. Washington: GPO, 1979.

———. ———. Senate. Committee on Commerce, Science, and Transportation. Tonnage Measurement Simplification Act. 96th Cong., 2nd sess. S. Report. 8852 to accompany H. R. 1197. Washington: GPO, 1980.

GOVERNMENT DOCUMENT WITH AUTHOR'S NAME

O'Neil, Tip. The Office and the Duties of the Speaker of the House of Representatives. U.S. 95th Cong., 2nd sess. H. Doc. 354. Washington: GPO, 1978.

When the author's name is known, you may list government documents either with the author's name first or with the government agency first.

Crain, Ben W., and Lloyd C. Atkinson. The European Monetary System, Problems, and Prospects: A Study. U.S. 96th Cong., 1st sess. Joint Economic Committee: Subcommittee on International Economics. Washington: GPO, 1979.

or:

United States. Cong. Joint Economic Committee: Subcommittee on International Economics. The European Monetary System, Problems, and Prospects: A Study. By Ben W. Crain and Lloyd C. Atkinson. 96th Cong., 1st sess. Washington: GPO, 1979.

EXECUTIVE BRANCH

United States. Central Intelligence Agency. National Basic Intelligence Factbook. Washington: GPO, 1980.

———. Department of State. United States Contributions to International Organizations. Washington: GPO, 1978.

———. President. Federal Personal Data Systems Subject to Privacy Act of 1974; Annual Report of the President. Pr. 39.11:978. Washington: GPO, 1978.

———. President's Committee on Employment of the Handicapped. Your New Mentally Retarded Worker. Pr. Ex. 1.10:M52/3. Washington: GPO, 1980.

JUDICIARY

Annual Report of the Immigration and Naturalization Service. Department of Justice, Immigration and Naturalization Service. Washington: GPO, 1978.

Emergency Petroleum Allocation Act of 1973. Statutes at Large, 87. Public Law 93-159 (1973).

Sears, Roebuck & Co. v. San Diego County District Council of Carpenters. 436 U.S. Reports, 76-150 (1977).

Don't underline legal references.

United States. Const. Art. I, sec. 9.

—. Supreme Court. Rules of the Supreme Court of the United States. Ju. 6.9:980. Washington: Supreme Court, 1980.

—. —. Heavy Lift Services, Inc., Petitioner v. National Labor Relations Board. No. 79-1655. Facts and Opinion, 607 F 2d 1121. 6 Oct. 1980.

United States v. Whitmire. 595 F. 2d 1303 (5th Cir. 1979).

United Steel Workers v. Weber. 47 U.S.L.W. 4852 (1979).

OTHER DOCUMENTS

Great Britain. Parliament. House of Commons. Report of the Select Committee on Police. New York: Arno Press & The New York Times, 1971.

New York City. The Scott Commission. Restructuring the Government of New York City. New York: Praeger, 1972.

New York State. Constitution. Amended to Jan. 1, 1970. New York: Department of State, 1970.

—. State Historical Association. History of the State of New York. 10 vols. New York: Columbia UP, 1933-1937.

—. New York State: A Citizen's Handbook. 3rd ed. New York: League of Women Voters of New York State, 1968.

OTHER SOURCES

ERIC DOCUMENTS

Heard, G. C., and L. D. Stokes. Psycho-Cultural Considerations in Black Students' Written Language Use: A Case Study, 1975. ERIC ED 137 815.

Perl, Sondra. The Composing Processes of Unskilled Writers at the College Level. Paper presented at the annual meeting of the Conference on College Composition and Communication, 1977. ERIC ED 147 819.

HANDOUT, MIMEOGRAPH, AND SO ON

Fraser, B. ''The Positioning of Conjoining Transformations in a Grammar.'' Mimeographed. Bedford, MA: Mitre Corporation, 1963.

Gleitman, L. ''Conjunction with and.'' Transformations and Discourse Analysis Projects, No. 40. Mimeographed. Philadelphia: U of Pennsylvania, 1961.

Olson, John L. ''Chronology of Renaissance Events.'' Handout for class in European History. Central Michigan U, 1979.

BULLETIN

Seashore, H. G., and J. H. Ricks, Jr. Text Service Bulletin, No. 39. New York: Psychological Corporation, 1950.

United States. Bureau of Labor Statistics. Exploring Careers, Bulletin 2001. Washington: GPO, 1979.

United States. Office of Education. Status of Education in the U.S. Bulletin No. 1. Washington: GPO, 1959.

PAMPHLET, LEAFLET

United States. Cong. House. The Working Congress. 94th Cong., 2nd sess. H. Doc. 94-623. H. Con. Res. 629. 9 Sept. 1976.

Women's Campaign Fund. Washington, DC, 1980.

LECTURE, SPEECH, PUBLIC ADDRESS

Haworth, Lorna H. ''Figuratively Speaking.'' Annual Meeting, National Council of Teachers of English, New York, 25 Nov. 1977.

Johnson, D. M. ''Problem-Solving Processes.'' Paper read at American Psychological Association, St. Louis, Sept. 1962.

POEM PUBLISHED SEPARATELY

Khayyam, Omar. The Rubaiyat. Trans. Edward Fitzgerald. Illus. Jeff Hill. Mount Vernon, NY: Peter Pauper Press, n.d.

See note to Beowulf entry, page 164.

POEM IN A COLLECTION

Yeats, William Butler. ''Crazy Jane Talks with the Bishop.'' In The Pocket Book of Modern Verse. Rev. ed. Ed. by Oscar Williams. New York: Washington Square, 1958. 195.

Note the treatment of the page number; no colon is used, nor is the page number introduced with "p." or "pp." or anything else.

FILM

Ashby, Hal, dir. Coming Home. With Jane Fonda, Jon Voight, and Bruce Dern. United Artists, 1978.

PLAY, PERFORMANCE

Lindsay-Hoagg, Michael, dir. Whose Life Is It Anyway? By Brian Clark. With Tom Conti and Jean Marsh. Trafalgar Theater, New York. 19 Apr. 1979.

PLAY, PUBLISHED SEPARATELY

Miller, Arthur. The Crucible. New York: Viking, 1953.

PLAY, IN A COLLECTION

O'Neill, Eugene. Mourning Becomes Electra. In Three Plays of Eugene O'Neill. New York: Vintage, n.d.

MUSICAL PERFORMANCE

Levine, James, cond. Parsifal. With Jon Vickers and Christa Ludwig. The Metropolitan Opera Company of New York. Metropolitan Opera House, New York. 16 Apr. 179.

MUSICAL COMPOSITION

Beethoven, Ludwig van. Symphony No. 7 in A, op. 92.

RADIO OR TELEVISION PERFORMANCE

''TV or Not TV.'' Commentator, Bill Moyers. Bill Moyers's Journal. PBS, 23 Apr. 1979.

TRANSCRIPT

''Joseph Campbell: Myths to Live by—Part I.'' Transcript of Bill Moyers's Journal, PBS, 17 Apr. 1981, New York: Educational Broadcasting Company, 1981.

RECORDING: ALBUM, TAPE CARTRIDGE, CASSETTE

Caruso, Enrico. The Best of Caruso. RCA Victor, LM-6056, 1958.

Horowitz, Vladimir. New Recordings of Chopin. Columbia TC8 Stereo Cartridge, MA 32932, 1974.

Williams, John, comp. and cond. Star Wars. London Symphony Orchestra. Original Sound Track. 20th Century Cassette, C2-541, 1977.

INDIVIDUAL SELECTION FROM ALBUM, CARTRIDGE, CASSETTE

Denver, John. ''Looking for Space.'' Windsong. RCA Album, Stereo APLi-1183, 1975.

Hamlisch, Marvin. ''Maple Leaf Rag.'' The Entertainer. MCA Cassette, MCAC-2115, 1974.

Kristofferson, Kris. ''Crippled Cow.'' In Barbra Streisand and Kris Kristofferson, A Star Is Born. Columbia TC8 Stereo Cartridge, JSA 34403, 1976.

WORK OF ART

Aphrodite (the ''Venus de Milo''). The Louvre, Paris.

Titian [Tiziano Vecellio]. Venus and Adonis. National Gallery of Art, Washington.

LETTER, PERSONAL

Easterly, K. T. Letter to the author. 6 Dec. 1979.

LETTER(S), PUBLISHED

Van Gogh, Vincent. The Complete Letters of Vincent Van Gogh. 3 vols. Greenwich, CT.: New York Graphic Society, 1958.

Identify the state as well as the city only if there is a possible ambiguity. Use ZIP code abbreviations for states. Supply names of foreign countries, Canadian provinces if necessary.

Van Gogh, Vincent, ''To Theo.'' Letter 358 in The Complete Letters of Vincent Van Gogh. Vol. II. Greenwich, CT: New York Graphic Society, 1958. 265–69.

Use the date of a letter, when it's present, along with any identifying number. If there is an editor, add (for example) "Ed. George Smith." after the title of the collection.

To cite more than one letter from a collection, give a reference to the collection in your bibliography and cite the various letters in your paper itself (Van Gogh 265–69).

PERSONAL INTERVIEW

Fonda, Jane. Personal interview. 10 Oct. 1978.

TELEPHONE INTERVIEW

Lamport, Loise. Telephone interview. 15 Jan. 1981.

PUBLISHED INTERVIEW

Greider, William. ''The Education of David Stockman,'' The Atlantic Monthly, Dec. 1981. 27–54.

If there is no title for the interview, add "Interview." after the interviewer's name.

COLLECTION: LETTERS, MANUSCRIPTS, PAPERS

Shaw, Charles. Papers. Archives of American Art. Smithsonian Institution.

ITEM IN COLLECTION

Mencken, H. L. Letter to Charles Shaw. Charles Shaw Papers. Archives of American Art, Smithsonian Institution.

Twain, Mark. Notebook 32, ts. Mark Twain Papers. U of California, Berkeley.

ILLUSTRATION: DRAWING, PHOTOGRAPH, TRANSPARENCY

da Vinci, Leonardo. Sketch of the hanged Baroncelli (enlarged). 1479. In Robert Wallace and the Editors of Time-Life Books. The World of Leonardo, 1452-1519. Alexandria, VA.: Time-Life, 1966.

''Globe Theater.'' Transparency. The Perfection Form Company, FS95134, 1981.

''Gross Anatomy: Principal Parts of the Human Body.'' Anatomical Chromographs (views 1-14). In Encyclopedia Britannica, 1970 ed.

Photograph from the Zapruder film. Commission Exhibit No, 902. In Report of the Warren Commission on the Assassination of President Kennedy. The New York Times ed. New York: McGraw, 1964.

CHART, TABLE, DIAGRAM

Broom, H. N., and Justice G. Longenecker. ''Chart of U.S. Consumer Credit for Selected Years.'' In Small Business Management. 5th ed. Cincinnati: South-Western, 1979.

Hawkins, Gerald S., with John B. White. Stonehenge computer; schematic plan. Stonehenge Decoded. Garden City, NY: Doubleday, 1965.

Porter, Sylvia. ''Savings Time Table.'' In Sylvia Porter's Money Book. Garden City, NY: Doubleday, 1975.

MASTER'S THESIS

Byers, J. L. ''An Investigation of the Goal Patterns of Academically Successful and Unsuccessful Children in a United States History Class.'' Master's Thesis, U of Wisconsin, 1958.

UNPUBLISHED PAPER

Smock, Erin. ''Proposal for a Pilot-test of Selected Procedures in Performance Measurement.'' Unpublished paper. Distributed at National Conference on Testing, Washington, 1981.

ACTIVITY 29 Edit the following MLA Works Cited page. Correct any errors of form or violations of MLA guidelines. Some entries may not contain any errors.

Kurt Anderson. ''Miami's new days of rage.'' Time, 10 Jan. 1983, pg. 20–21.

Barron, James. ''Officer Shot Dead, 2 Wounded.'' New York Times, February 15, 1984, Sec. A. p. 1, col. 2.

Brown, Michael F. ''Use of Deadly Force by Patrol Officers: Training Implications.'' Journal of Police Science and Administration Vol. 12 (1984), 133–140.

Bridgewater, Carol Austin. '''Hill Street' Burnout Blues.'' Psychology Today May 1984: 70.

Cullen, Francis T., Bruce G. Link, Lawrence F. Travis, II, and Terrence Lemming. ''Paradox in Policing: A Note on Perceptions of Dangers.'' Journal of Police Science and Administration, 11 (March, 1983), 457–465.

Meredith, Nikki. ''A Better Way in San Jose.'' Psychology Today, May 1984, p. 25.

Meredith, Nikki. ''Attacking the Roots of Police Violence.'' Psychology Today May 1984: 21–6.

Anonymous. ''Police Brutality Hot Issue Again.'' U.S. News and World Report, March 26, 1984, p. 13.

Vernon B. Wherry, ''Research Into The Use Of Deadly Force,'' The Police Chief, October 1981, pp. 44–49.

Lambert, Richard D., and Allan Heston, eds. The Police and Violence. Philadelphia: American Academy of Political and Social Science, 1980.

NAME AND DATE METHOD OF DOCUMENTATION: APA STYLE

An alternative style of documentation is the name and date method, based on the style sheet of the American Psychological Association. This method is often used in science, education, and business. The name-and-date method does away with footnotes and endnotes. Some researchers feel that the citations in the text are helpful because the reader can identify authorities and dates of research immediately; there is no need to look for them at the foot of the page or end of the paper. If you plan to cite whole books and articles without quoting directly from them, this may be the preferred form of documentation:

> It has been demonstrated that chemical substitutes eventually become more expensive than the natural substances they replace. (Lloyd, 1975; Meger, 1987; Royce, 1980).

Such documentation means that the information can be found in all three sources, suggesting this information is well known or at least readily available to researchers. The fact that it's well known is made clear through the form of the documentation. On the other hand, if there are many such citations on a page, they can become intrusive, sacrificing readability for documentation:

> The experiment was ''inherently flawed'' (Clay, 1977, p. 12) and projects based on its results soon consumed millions of wasted dollars (Evers, 1978, pp. 19–20) and sidetracked dozens of projects for years afterward (Hayne, 1977; North, 1977; Wenrow, 1980).

You can relieve the clutter of citations by using an option: give some references in the text itself, without parentheses. See Option, below.

Guidelines for Citations in the Text

At the most appropriate place in the text, give the author's name, followed by comma, space, and date.

> This fact has been known for at least a hundred years (see Frankel, 1879).
>
> Koko signs to herself while looking at pictures, names things she can see without requesting them, exhibits, ''naming behavior'' (Patterson, 1978).

Option As an alternative, the author's name can be given in the text, with only the date supplied in parentheses.

> According to Patterson (1978), Koko signs to herself while looking at pictures, names things she can see, without requesting them, exhibits ''naming behavior.''

Both the name and date can be incorporated into your text. In this case, no parenthetical information is required.

> In a 1978 report, Patterson says Koko signs to herself while looking at pictures, names things she can see without requesting them, exhibits ''naming behavior.''

You must give a full description in your reference list for a note like this, just as if it were in parentheses.

Extended Discussion

You needn't keep citing a source as long as no intervening source is introduced. If the discussion is extremely lengthy, you may give just the name (Mall) where appropriate to remind the reader. Nevertheless, in general, it's better to give additional citations than to risk ambiguous documentation.

Two or More Authors

For two authors, always give both authors' names (last names only). For three to five authors, give all authors' names in the first citation; thereafter use only the first author's name with "et al." For six or more authors, give only the first author's name and "et al." for all citations. However, in your Reference List at the end of your paper, always give all authors' names, no matter how many.

FIRST CITATION

`(Savage-Rumbaugh, Pate, Lawson, Smith, & Rosenbaum, 1983)`

SUBSEQUENT CITATIONS

`(Savage-Rumbaugh, et al., 1983)`

If the names are given in the text of your paper, without parentheses, use "and" instead of an ampersand: Savage-Rumbaugh, Pate, Lawson, Smith, *and* Rosenbaum (1983).

NO AUTHOR

If there is no author's name, use either the title or an abbreviated (but recognizable) form of the title: (*New Technology for Libraries*, 1979).

LONG NAME

Lengthy names of groups or committees may be shortened (NCTE, 1981) so long as the reader will be able to recognize the name in the bibliography: The National Council of Teachers of English.

TWO AUTHORS WITH SAME NAME

Use initials to identify authors with the same last name: (Kennedy, G., 1963), (Kennedy, J., 1964).

SAME AUTHOR, SAME YEAR

Two or more works by the same author published in the same year should be further identified with lower-case letters in parentheses: (a), (b).

IN YOUR PAPER

. . . (Christensen, 1963a, 1963b) . . .

IN REFERENCE LIST

Christensen, F. (1963a). Generative rhetoric of the sentence. College Composition and Communication, 14, 155-161.

Christensen, F. (1963b). Notes toward a new rhetoric. College English, 25, 7-18.

MULTIPLE REFERENCES

If there is more than one reference, follow the order in the reference list, usually alphabetical: (Allan, 1971; Mailer, 1979; Pressman, 1977). References to the same author also follow the order in the reference list—chronological rather than alphabetical: (Kennedy, 1963, 1972, 1980).

PAGE NUMBERS

Specific pages, quoted material, paraphrases, anything that would require a page number in MLA style (q.v.) will also require one in the name and date style: (Burger, 1981, p. 12). Note that this style uses "p." to cite page and "pp." for pages. Give all page numbers in full; do not shorten inclusive numbers: (pp. 150–157, not 150–57).

The Reference List in APA Style

The reference list contains all the sources used in your paper, and only the sources you actually cite. When you use the name-and-date style, the list is always called "References" (not "Bibliography," which may include sources not cited). The purpose of the reference list is to help other researchers find the materials you have used. You must provide complete and accurate information taken from the title page of a book (never from the cover). However, use only the initials of authors' first and middle names, even if spelled in full on the title page. Information for newspaper, magazine, or journal articles should be given as it appears in the publication and not as abbreviated in tables of contents or periodical indexes.

There are so many variations in references that it's impossible to give examples for all of them. Those given here are the ones most likely to be needed by students. Others may be readily inferred from those given here or from those listed in the MLA section. Occasionally you may find a source for which there is no exact documentation model; in that case you must invent a reasonable application of the general principles discussed below.

Order of Information

Author(s), if any (use last name and initials only of all names)
Date of publication
Title of book, article, report, and so on
Publication information: Books require place and name of publisher. Journals, magazines, and newspapers: Give volume and issue numbers where appropriate and page numbers.

Punctuation

Periods are placed after the four main divisions: author, date, title, and publication information. Within these divisions, the comma is the most frequently used mark of separation. Explanatory information is usually set off in parentheses.

Capitalization

Capitalize the first word and all significant words in journal and magazine titles. For any other titles (books, articles, essays) capitalize only the first word and proper names. The first word of a subtitle after a colon or dash should be capitalized.

Special Treatment of Titles

Underline the titles of books, journals, magazines, and newspapers. Underline volume numbers in journal references. Don't put quotation marks around titles. The only exception to this rule occurs when one title appears inside another:

```
Lance, P. (1981). An analysis of Poe's ''The Masque of the Red
     Death.'' American Poetry, 19(3), 36-46.
```

Roman Numerals

Convert roman numerals to arabic (volume 4 instead of volume IV), unless the source material is numbered that way (for example, a preface) or the roman numeral is part of the title of a source. See chart of roman numerals, page 243.

EXAMPLES

Author(s) name(s), last name first (note that all author's names are inverted, unlike MLA, which inverts only the first listed author's name), initials only for first and middle names, even if full name is used on title page.

Ampersand in place of *and* (but use *and* if referring to authors in your paper: "Downie *and* Heath show the formula . . .").

Two spaces between name and date. (Two spaces after any end mark.) Period after initial serves as end mark.

Date of publication in parentheses (frequently given as "copyright" on title page), followed by period.

Book title underlined. Wording as given on title page, not book cover.

Capitalize only first word, proper names, and proper adjectives.

```
Downie, N. M., & Heath, R. W.  (1965). Basic statistical methods
     (2nd ed.). New York: Harper and Row.
```

Indent all subsequent lines five spaces.

Explanatory, clarifying information given in parentheses.

Period ending book title (and explanatory material).

Place of publication, if major city; give state, province, nation also if city may be ambiguous (London, Ontario). List only first city if several are given on title page.

Colon and space between place and name of publisher.

Publisher's name in slightly shortened form. Drop redundant information like "Co.," "Inc.," etc.

Period at end of reference.

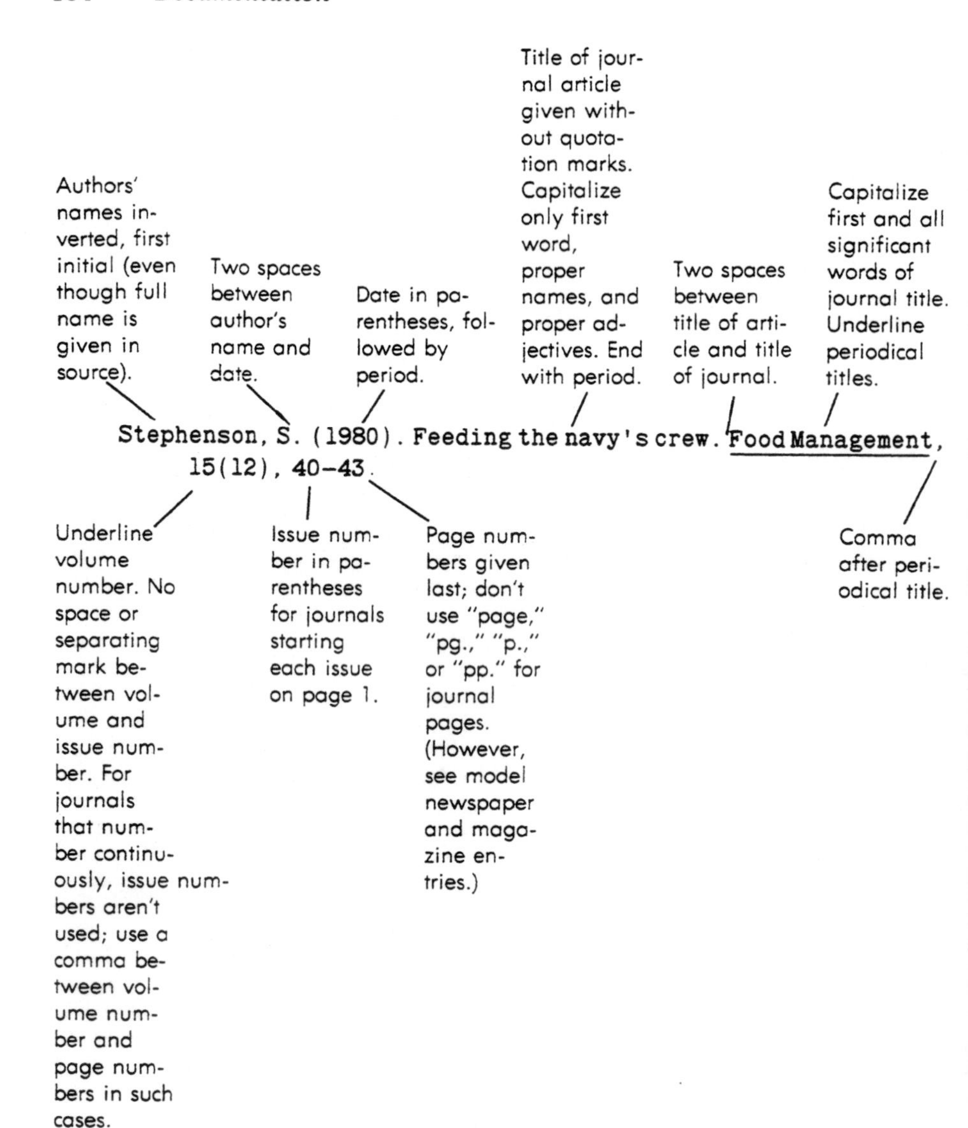

REFERENCE LIST MODELS IN APA STYLE: BOOKS

ONE AUTHOR

Allan, C. M. (1971). The theory of taxation. Baltimore: Penguin.

Note initials in author's name, capitalization in title, shortened name for publisher.

Mailer, N. (1979). The executioner's song. Boston: Little, Brown.

Pressman, R. S. (1977). Numerical control and computer-aided manufacturing. New York: John Wiley.

Wolfe, T. (1980). The right stuff. New York: Bantam.

Note alphabetical order for items in reference list.

MORE THAN ONE BOOK BY SAME AUTHOR

Treat these books as if they were written by different authors with the same name. Don't use a string of hyphens. Chronological order is required.

Kennedy, G. A. (1963). The art of persuasian in Greece. Princeton, NJ: Princeton University Press.

Kennedy, G. A. (1972). The art of rhetoric in the roman world, 300 B.C.–A.D. 300. Princeton, NJ: Princeton University Press.

Kennedy, G. A. (1980). Classical rhetoric and its Christian and secular tradition from ancient to modern times. Chapel Hill: University of North Carolina Press.

Identify most states except those very well known or those identified in the publisher's name. Spell out in full all university press names (do not abbreviate).

MORE THAN ONE BOOK BY SAME AUTHOR, WITH CO-AUTHOR

Postman, N. (1979). Teaching as a conserving activity. New York: Delacorte.

Postman, N., & Weingartner, C. (1969). Teaching as a subversive activity. New York: Delacorte.

Co-authored books are subsequent to books with one author, regardless of the date.

TWO OR MORE AUTHORS

Downie, N. M., & Heath, R. W. (1965). Basic statistical methods (2nd ed.). New York: Harper & Row.

Give explanatory information in parentheses. Invert all authors' names.

Field, F., Meacher, M., & Pond, C. (1977). To him who hath: A study of poverty and taxation. New York: Penguin.

You must give all authors' names, no matter how many. Don't use "et al." in your Reference List (but you may use it in your parenthetic citations after giving all the names once.)

Sagan, C., Drake, F. D., Druyan, A., Ferris, T., Lomberg, J., & Sagan, L. S. (1978). Murmurs of earth: The voyager interstellar record. New York: Random House.

MULTIPLE PUBLISHERS

Phelps, R. H. (1969). Mass violence in America: A history of Newgate of Connecticut. New York: Arno Press & the New York Times.

New technology for libraries: A layman's guide to reducing public library costs and improving service through scientific methods and tools. (1979). (A background paper for the White House Conference on Library and Information Services.) New York: National Citizens Emergency Committee to Save our Public Libraries; Washington: National Commission on Libraries and Information Science; American Library Association; Urban Libraries Council.

Groups, committees, and others sometimes publish unsigned documents—that is, without an author's name. Unsigned works aren't considered anonymous, which implies that the author's name is unknown. Such unsigned works are listed by title and cited in the text by an abbreviated but recognizable form of the title.

COMMITTEE OR GROUP AUTHOR

Committee on the National Interest. (1964). The national interest and the continuing education of teachers of English. Champaign, IL: National Council of Teachers of English.

Group authors and unsigned works are alphabetized according to the first significant word in the name. Don't alphabetize under *A, An,* or *The*. *The Zuider Report* would be alphabetized as if it started with *Zuider*.

COMMITTEE OR GROUP, AUTHOR NAMED IN TITLE

Report of the Warren commission on the assassination of president Kennedy (New York Times Edition). (1964). New York: McGraw-Hill.

COMMITTEE OR GROUP, AUTHOR SAME AS PUBLISHER

Canadian Tax Foundation. (1979). Provincial and municipal finances. Toronto: Author.

BOOK WITH EDITOR(S)

Hartman, F. (Ed.). (1973). World in crisis: Readings in international relations (4th ed.). New York: Macmillan.

Bullock, A., & Stallybrass. O. (Eds.). (1977). The Harper dictionary of modern thought. New York: Harper & Row.

CHAPTER OR SECTION OF BOOK BY ONE AUTHOR

Asimov, I. (1972). What is science? Asimov's guide to science. New York: Basic Books.

Page numbers, if any, would appear in the parenthetic citation in your paper, not in the reference list.

CHAPTER OR SECTION IN AN EDITED WORK

Kline, M. (1972). The meaning of mathematics. In R. Thruelsen & J. Kobler (Eds.), Adventures of the mind, 2nd series. (pp. 73-98). New York: Vintage.

Note that editors' names aren't inverted here. (See Book with Editor(s), above.) Page numbers are required.

TRANSLATION

Dostoyevsky, F. (1950). Crime and punishment (C. Garnett, Trans.). New York: Modern Library.

MULTI-VOLUME WORK

Churchill, W. S. (1948-1953). The second world war (6 vols.). Boston: Houghton Mifflin.

PART(S) OF MULTI-VOLUME WORK

Churchill, W. S. (1953). Triumph and tragedy (Vol. 1 of The second world war). Boston: Houghton Mifflin.

Durant, W. (1953). The renaissance (Part 5 of The story of civilization). New York: Simon & Schuster.

Steffens, L. (1958). The autobiography of Lincoln Steffens (Vol. 2). New York: Harcourt, Brace, & World.

Note that roman numerals have been replaced with arabic volume numbers. Only if the volume number is part of the title itself should you allow it to remain in roman type.

ANONYMOUS, AUTHOR UNKNOWN

Beowulf. (1967). (W. E. Leonard, Trans., L. Ward, Illus.). Norwalk, CT: Heritage Press.

The book of Kells: Reproductions from the manuscript in Trinity College Dublin. (1974). New York: Alfred A. Knopf.

Only if the word "anonymous" appears on the title page of the book should it appear in your references. It would then be capitalized and treated as the name of the author.

ANONYMOUS, NAME SUPPLIED

The singular adventures of Baron Munchausen [R. Raspe, et al., auth.]. (1952). Norwalk, CT: Heritage Press.

PSEUDONYM (PEN NAME), NAME SUPPLIED

Twain, M. (Pseud. of S. L. Clemens). (1959). The autobiography of Mark Twain (C. Neider, Ed.). New York: Harper & Brothers.

SERIES

Brée, G. (Ed.) (1962). Camus, a collection of critical essays. Twentieth Century Views. Englewood Cliffs, NJ: Prentice-Hall.

Peyre, H. (1962). Camus the pagan. In G. Brée (Ed.), Camus, a collection of critical essays. Twentieth Century Views. (pp. 65–70). Englewood Cliffs, NJ: Prentice-Hall.

MONOGRAPH

Coleman, J. S. (1961). Social climates in high schools (Cooperative Research Monographs. No. 4). Washington, DC: U.S. Office of Education.

REPRINT OF OLDER WORK

Agee, J., & Evans, W. (1966). Let us now praise famous men. New York: Ballantine. (Originally published, 1939.)

Parenthetic references to this work in your paper should give both dates: (Agee & Evans, 1939/1966).

EDITION

Gunther, E. (1973). Ethnobotany of Western Washington: The knowledge and use of indigenous plants by native Americans (rev. ed.). Seattle: University of Washington Press.

INTRODUCTION, PREFACE, FOREWORD

Cowley, M. (Ed.). (1959). How writers write (introd.). Writers at work: The Paris Review interviews. New York: Viking.

BIBLE

The Holy Bible (authorized King James version). New York: Harper & Brothers, n.d.

The New English Bible with Apocrypha. (1970). Oxford University Press; Cambridge University Press.

Don't underline titles of versions of the Bible.

ENCYCLOPEDIA, ALPHABETICALLY ARRANGED WORK

Archibald, K. (1976). Cerebral palsy. Encyclopedia Americana.

Chickamauga Dam. (1976). Encyclopedia Americana.

DISSERTATION, UNPUBLISHED

Kitzhaber, A. R. (1953). Rhetoric in American colleges 1850-1900. Unpublished doctoral dissertation, University of Washington.

ARTICLE FROM DISSERTATION ABSTRACTS OR DISSERTATION ABSTRACTS INTERNATIONAL

Hazen, C. L. (1973). The relative effectiveness of two methodologies in the development of composition skills in college freshman English (Doctoral dissertation, North Texas State University). Dissertation Abstracts International, 33, 4243A.

Note: "A" after a page reference indicates humanities, and "B" indicates sciences in *Dissertation Abstracts*. The title was changed to *Dissertation Abstracts International* starting with volume 30.

REFERENCE LIST MODELS IN APA STYLE: PERIODICALS

WEEKLY MAGAZINE ARTICLE

Raskin, A. H. (1980, August 25). A reporter at large: After Meany. The New Yorker, pp. 36–76.

MAGAZINE ARTICLE, NO AUTHOR GIVEN

Byrd of West Virginia: Fiddler in the Senate. (1978, January 23). Time, pp. 12–16.

MONTHLY MAGAZINE ARTICLE

Peirce, R. M. (1980, December). Your dollar is at a premium. The Nation's Business, pp. 92–100.

Nontechnical sources, like popular magazines, use "p." and "pp." for "page" and "pages." Note the capitalization of all significant words in the title of the periodical, but not the title of an article.

NEWSPAPER ARTICLE

Talbert, B. (1978, January 10). Why are we in such a downer? Detroit Free Press, sec. A, p. 9.

Give all page numbers separated by commas for discontinuous pages: pp. 3, 5, 9.

REVIEWS, SIGNED AND UNSIGNED

Aire, S. (1979). Review of Othello, performed by the Royal Shakespeare Society at Stratford, August 1, 1979. Plays and Players, 27, 23–24.

Note that volume numbers are underlined. Page numbers aren't preceded with "p." or "pp." when a volume number is given. (See "Professional, Technical, or Specialty Journal," below.)

Paperbacks: New and Noteworthy [Review of Let us now praise famous men, by James Agee and Walker Evans]. (1981, January 4). New York Times Book Review, p. 23.

EDITORIAL

Mr. Burger's case for prison reform [Editorial]. (1980, December 30). The Christian Science Monitor, p. 24.

LETTER TO THE EDITOR

Taylor, J. C. (1980, December). China revisited [Letter to the editor]. Life, p. 44.

PROFESSIONAL, TECHNICAL, OR SPECIALTY JOURNAL, EACH ISSUE STARTING WITH PAGE 1

Clarke, B. (1980). Notes on a no-win campaign. Columbia Journalism Review, 19(3), 36–40.

Note the issue number is in parentheses. Since each volume usually has several issues in it, readers need to know the issue number in order to find pages 36–40. There is no space between volume and issue numbers.

PROFESSIONAL, TECHNICAL, OR SPECIALTY JOURNAL, PAGES NUMBERED CONTINUOUSLY THROUGHOUT VOLUME

Blum, L., & Givant, S. (1981). Increasing the participation of women in fields that use mathematics. The American Mathematical Monthly, 87, 6–22.

REFERENCE LIST MODELS IN APA STYLE: GOVERNMENT PUBLICATIONS AND OTHER SOURCES

GOVERNMENT PUBLICATIONS

Agreement with Univ. of Texas for Lyndon B. Johnson presidential archival depository. (1965). (89th Cong., 1st sess., House Report 892), Washington, DC: U.S. Government Printing Office.

O'Neill, Tip. (1978). The office and duties of the speaker of the house of representatives (95th Cong., 2nd sess., House Document 354). Washington, DC: U.S. Government Printing Office.

Committee on Commerce, Science and Transportation. (1980). Tonnage measurement simplification act (96th Cong., 2nd sess., Senate Report 852 to accompany House Report 1197). Washington, DC: U.S. Government Printing Office.

Federal Personal data systems subject to privacy act of 1974: Annual report of the President (Pr 39. 11:978). (1978). Washington, DC: U.S. Government Printing Office.

Central Intelligence Agency. (1980). National basic intelligence factbook. Washington, DC: U.S. Government Printing Office.

Rules of the Supreme Court of the United States (Ju. 6.9:980). (1980). Washington, DC: Supreme Court.

Sears, Roebuck & Co. v. San Diego County District Council of Carpenters. (1977). U.S. Reports, 436, 76–750.

Economic Commission for Africa. (1973). African regional plan for the application of science and technology to development. New York: United Nations Publication.

ERIC DOCUMENTS

Perl, S. (1977). The composing processes of unskilled writers at the college level. Paper presented at the annual meeting of the Conference on College Composition and Communication. (ERIC Document Reproduction Service No. ED 147 819.)

ERIC is the Educational Resources Information Center. See page 49.

HANDOUT, PHOTOCOPY, MIMEOGRAH, DITTO, AND SO ON

Gleitman, L. (1961). Conjunctions with ''and.'' (Transformations and discourse analysis projects, no. 40). Unpublished manuscript, Philadelphia: University of Pennsylvania.

BULLETIN, REPORT

Office of Education. (1959). Status of education in the U.S. (Bulletin No. 1). Washington, DC: U.S. Government Printing Office.

Seashore, H. G., & Ricks, J. H. (1950). Test service bulletin (No. 39). New York: the Psychological Corporation.

Bar-Hillel, Y., Kasher, A., & Shamir, E. (1963). Measures of syntactic complexity (Report for the U.S. Office of Naval Research, Information Systems Branch). Washington, D.C.: U.S. Government Printing Office.

LECTURE, SPEECH, PUBLIC ADDRESS

Haworth, L. H. (1977, November 25). Figuratively speaking. Paper presented at the annual meeting of the National Council of Teachers of English, New York.

FILM

Ashby, H. (Director). (1978). Coming Home. [Film]. United Artists.

POEM PUBLISHED SEPARATELY

Khayyam, O. (n.d.). The rubaiyat. E. Fitzgerald (Trans.), J. Hill (Illus.). Mount Vernon, NY: Peter Pauper Press.

POEM IN A COLLECTION

Yeats, W. B. (1958). Crazy Jane talks with the bishop. In O. Williams (Ed.), The pocket book of modern verse (rev. ed.). New York: Washington Square Press.

PLAY, PERFORMANCE

Lindsay-Hoagg, M. (Director). (1979, April 19). Whose life is it anyway? [Play]. By Brian Clark. Trafalgar Theater, New York.

RADIO OR TELEVISION PERFORMANCE

Moyers, B. (Commentator). (1979, April 23). TV or not TV. [Television]. In Bill Moyers' Journal, PBS.

WORK OF ART

Aphrodite (the ''Venus de Milo''). The Louvre, Paris.

LETTER, PERSONAL

Easterly, K. T. (1979, December 6). Letter to author.

LETTER(S), PUBLISHED

Van Gogh, V. (1958). Letter to Theo (No. 358). In The complete letters of Vincent Van Gogh (Vol. 2). Greenwich, CT: New York Graphic Society.

PERSONAL INTERVIEW

Fonda, J. (1978, October 10). Personal interview.

ILLUSTRATION: DRAWING, PHOTOGRAPH, TRANSPARENCY

Gross Anatomy: Principal parts of the human body (Anatomical chromographs, views 1–14). (1970). In Encyclopedia Britannica.

Photograph from the Zapruder film (Commission Exhibit No. 902). (1964). In Report of the Warren Commission on the assassination of President Kennedy (New York Times ed.). New York: McGraw-Hill.

CHART, TABLE, DIAGRAM

Hawkins, G. S., & White, J. B. (1965). Stonehenge computer; schematic plan. In Stonehenge decoded. Garden City. NY: Doubleday.

UNPUBLISHED PAPER

Smock, E. (n.d.). *Proposal for a pilot-test of selected procedures in performance measurement*. Unpublished paper distributed at National Conference on Testing, Washington, DC, 1981.

ACTIVITY 30 Edit the following APA References page. Correct any mistakes in form, violations of APA guidelines. Some items may not need editing.

Alexander, Shana. ''Simple Question of Rape.'' *Newsweek*, p. 110, Oct. 28, 1974.

Burgess, A. W., & L. L. Holmstrom. Coping behavior of the rape victim. *American Journal of Psychiatry*, Vol. 133, (1976), pgs. 413–417.

Clark, L., & Lewis, D. (1977). *Rape: The Price of Coercive Sexuality*. Toronto: Women's Press.

Cohen, M., Garofalo, R., Boucher, J., and Seghorn, T. (1971). The psychology of rapists. *Seminars in psychiatry*, *3*, 307–327.

Arnold, P. (1975). *Lady beware*. Garden City, NY: Doubleday.

Fetter, Ann and Sanford, Linda. Defending yourself against rape. *Cosmopolitan*, Feb. 1981, pp. 106–9.

——. *In Defense of Ourselves*. Garden City, New York: Doubleday and Company, Inc., 1979.

James Selkin, ''Protecting Personal Space: Victim and Resister Reactions to Assaultive Rape.'' *Journal of Community Psychology*, Vol. 6 NO. 3 (1978), 263–68.

Sobel, Dava. ''Similar Motivation Seen In Rape Of Wife And Of A Stranger. *New York Times*, Feb. 5, 1980, Sec. C, p. 1.

TRADITIONAL FOOTNOTES AND ENDNOTES

Some disciplines continue to use the traditional documentation style, which calls for numbers in your text (raised half a line, like this[8]) and matching footnotes at the foot (bottom) of the page or collected as endnotes on a separate page or pages at the end of your paper. If footnotes are to be used, triple-space between your last line and the first note. Number the notes consecutively throughout your paper.

However, most researchers prefer endnotes, and for college work they are usually preferable to footnotes. Endnotes give the same information in the same order as footnotes, and they are numbered consecutively throughout your paper. Endnotes are collected at the end of the paper on a separate page or pages. You will find that it is considerably easier to use endnotes than to allow room on each page for footnotes. See a model paper in the Appendix.

FOOTNOTE/ENDNOTE DOCUMENTATION FORMS

The note and bibliography models below are based on *The Chicago Manual of Style*, 13th edition.

BOOK, ONE AUTHOR

Note Form

1. Norman Mailer, The Executioner's Song (Boston, Little, Brown, 1979), 597–98.

The Chicago Manual prefers to itemize each entry with a number followed by a period at the left margin. But you may, if you prefer, use the style in which the first line is indented, and each item is identified with a superior number (raised half a line):

[1] Norman Mailer, The Executioner's Song (Boston: Little, Brown, 1979), 597–98.

Bibliography Form

Mailer, Norman. The Executioner's Song. Boston: Little, Brown, 1979.

SUBSEQUENT REFERENCE TO AN AUTHOR

The Latin notes for subsequent references—*ibid.* and *op. cit.*—are passing out of style. If you use them, be sure to use them properly (see the list of abbreviations at the end of Chapter six). The preferred style for subsequent references is an abbreviated form of the first reference. A second reference to Mailer's book would use his last name and a page number:

2. Mailer, 47.

If you have more than one book by the same author, second references should contain the author's name and the title of the work. For example, note two works by Kennedy, and a second reference for one of them:

MORE THAN ONE BOOK BY SAME AUTHOR

Note Forms

3. George A. Kennedy, The Art of Persuasion in Greece (Princeton, N.J.: Princeton Univ. Press, 1963) 97.

4. George A. Kennedy, The Art of Rhetoric in the Roman World, 300 B.C.–A.D. 300 (Princeton, N.J.: Princeton University Press, 1972), 215.

5. Kennedy, The Art of Persuasion in Greece, 98.

Do not repeat an author's name with subsequent books. A string of five hyphens takes the place of the author's name. Note optional chronological order for two or more books by same author.

Bibliography Forms

Kennedy, George A. The Art of Persuasion in Greece. Princeton, N.J.: Princeton University Press, 1963.

———. The Art of Rhetoric in the Roman World, 300 B.C.–A.D. 300. Princeton, N.J.: Princeton University Press, 1972.

———. Classical Rhetoric and Its Christian and Secular Tradition from Ancient to Modern Times. Chapel Hill: University of North Carolina Press, 1980.

If you have two authors with the same last name, give both first and last names in subsequent references:

6. John F. Kennedy, 219.

TWO OR MORE AUTHORS

Note Forms

7. N. M. Downie and R. W. Heath. Basic Statistical Methods, 2nd ed. (New York: Harper & Row, 1965), 22.

8. Frank Field, Molly Meacher, and Chris Pond. To Him Who Hath (New York: Penguin Books, 1977), 80.

Note: The subtitle is optional in endnotes and footnotes. (However, it is required in the bibliography.)

For more than three authors, use "et al." ("and others") in your notes, but you should list all the names in the bibliography.

9. Carl Sagan, et al., Murmurs of Earth (New York: Random House, 1978), 17.

Bibliography Forms

Note that only the first author's name is inverted. The names of second and third authors are given in normal order.

Downie, N. M., and R. W. Heath. *Basic Statistical Methods*. 2nd ed. New York: Harper, 1965.

Field, Frank, Molly Meacher, and Chris Pond. *To Him Who Hath: A Study of Poverty and Taxation*. New York: Penguin, 1977.

Sagan, Carl, F. D. Drake, Ann Druyan, Timothy Ferris, Jon Lomberg, and Linda Salzman Sagan. *Murmurs of Earth: The Voyager Interstellar Record*. New York: Random House, 1978.

BOOK WITH EDITOR(S)

Note Forms

11. Frederick H. Hartman, ed. *World in Crisis*, 4th ed. (New York: Macmillan, 1973), 51.

12. Allan Bullock and Oliver Stallybras, eds. *The Harper Dictionary of Modern Thought* (New York: Harper, 1977), 173.

Bibliography Forms

Hartman, Frederick H., ed. *World in Crisis: Readings in International Relations*, 4th ed. New York: Macmillan, 1973.

Bullock, Allan, and Oliver Stallybras, eds. *The Harper Dictionary of Modern Thought*. New York: Harper, 1977.

ESSAY, CHAPTER OR SECTION IN AN EDITED WORK

Note Form

13. Morris Kline, ''The Meaning of Mathematics,'' in *Adventures of the Mind*, eds. Richard Thruelsen and John Kobler, 2nd Series (New York: Vintage Books, 1972), 77–91.

Bibliography Form

Kline, Morris. ''The Meaning of Mathematics.'' In *Adventures of the Mind*, 2nd Series, edited by Richard Thruelsen and John Kobler, 77–91. New York: Vintage, 1972.

WEEKLY MAGAZINE ARTICLE

Note Form

14. Lance Morrow, ''It's Time to Ban Handguns,'' *Time*, 13 Apr. 1981, 51.

Bibliography Form

Morrow, Lance. ''It's Time to Ban Handguns.'' Time, 13 Apr. 1981, 51.

Abbreviate months except May, June, July. Note that the day precedes the month, and no comma is needed to separate the month from the year.

MAGAZINE ARTICLE, NO AUTHOR GIVEN

Note Form

15. ''Byrd of West Virginia: Fiddler in the Senate,'' Time, 23 Jan. 1978, 14.

Bibliography Form

''Byrd of West Virginia: Fiddler in the Senate.'' Time, 23 Jan. 1978, 12-16.

SECOND REFERENCE FOR ARTICLE WITHOUT AUTHOR

16. ''Byrd of West Virginia,'' 13.

NEWSPAPER ARTICLE

Note Form

17. Bob Talbert, ''Why Are We In Such a Downer?'' Detroit Free Press, 10 Jan. 1978, Sec. A, p. 9.

Note: "p." or "pp." may be used in newspaper notes to prevent confusion.

Bibliography Form

Talbert, Bob. ''Why Are We in Such a Downer?'' Detroit Free Press, 10 Jan. 1978, sec. A, p. 9.

NEWSPAPER ARTICLE, UNSIGNED

Note Form

18. ''Latest Bid from Iran Won't Free Hostages,'' Detroit Free Press, 28 Dec. 1980, Sec. A, p. 9.

Bibliography Form

''Latest Bid from Iran Won't Free Hostages.'' Detroit Free Press, 28 Dec. 1980, Sec. A, p. 1, p. 9.

PROFESSIONAL, TECHNICAL, OR SPECIALTY JOURNAL, EACH ISSUE STARTING WITH PAGE 1

Note Form

19. Blair Clark, ''Notes on a No-Win Campaign,'' Columbia Journalism Review 19, no. 3 (1980): 39.

20. ''SFPA Hears Grim Forecast for World Economy, Hope for U.S.,'' Forest Industries 107, no. 13 (1980): 13.

You may give the full date instead of an issue number: 19 (Nov. 1980): 39.

Bibliography Forms

Clark, Blair. ''Notes on a No-Win Campaign.'' Columbia Journalism Review 19, no. 3 (1980): 36–40.

''SFPA Hears Grim Forecast for World Economy, Hope for U.S.'' Forest Industries 107, no. 13 (1980): 11–13.

PROFESSIONAL, TECHNICAL, OR SPECIALTY JOURNAL, PAGES NUMBERED CONTINUOUSLY THROUGHOUT VOLUME

Note Forms

21. Lenore Blum and Steven Givant, ''Increasing the Participation of Women in Fields That Use Mathematics,'' The American Mathematics Monthly, 87 (1981): 18.

22. Masao Furukawa, ''Wall Temperatures and Solar Deflection of Cylindrical Structures in Space,'' Journal of Spacecraft and Rockets 17 (1980): 501.

23. Bart McDowell, ''The Aztecs,'' National Geographic 158 (1980): 751.

Bibliography Forms

Blum, Lenore, and Steven Givant. ''Increasing the Participation of Women in Fields That Use Mathematics.'' The American Mathematics Monthly 87 (1981): 6–22.

Furukawa, Masao. ''Wall Temperatures and Solar Deflection of Cylindrical Structures in Space.'' Journal of Spacecraft and Rockets 17 (1980): 501–14.

McDowell, Bart. ''The Aztecs.'' National Geographic 158 (1980): 704–51.

POEM IN A COLLECTION

Note Form

24. William Butler Yeats, ''Crazy Jane Talks with the Bishop,'' in The Pocket Book of Modern Verse, rev. ed., edited by Oscar Williams, 195. (New York: Washington Square Press, 1958).

Bibliography Form

Yeats, William Butler. ''Crazy Jane Talks with the Bishop.'' In The Pocket Book of Modern Verse. Rev. ed., edited by Oscar Williams, 195. New York: Washington Square Press, 1958.

LINE IN A POEM

Note Form

25. Dante, ''Inferno,'' canto 14, line 4.

"Inferno" is the first book of *The Divine Comedy.*

PLAY, PUBLISHED SEPARATELY

Note Form

26. Arthur Miller, The Crucible (New York: Viking Press, 1953), Act 3.

Bibliography Form

Miller, Arthur. The Crucible. New York: Viking, 1953.

LINE IN A PLAY

Note Form

27. Shakespeare, Cymbeline, 1.5. 19–20.

The reference is to Act one, Scene five, lines 19–20. In general, use only arabic numerals, not roman.

ACTIVITY 31 Edit the following Endnotes page. Correct any errors in form.

A. Van Allen, ''NASA and the planetary imperative.'' Sky and Telescope, 64 (1982), pg. 320–322.

Clines, Francis X. ''NASA Gets Gift of $60,000 to Continue Viking Mission,'' New York Times, January 8, 1981, Sec. A. p. 20, col. 1.

Eberhart, Jonathan. A New Path to the Planets. Science News, Vol. 123 (1983), 250, 251.

———————. (1980) ''Mars: Radar Hints At Liquid Water.'' Science News, Vol. 117, p. 230, 231.

Kobrick, Michael. ''Topography of the Terrestrial Planets.'' Astronomy, May 1982, pp. 18–22.

McKay, Christopher P., and Stoker, Carol. Mission to Mars: The case for a settlement, Technology Review, Nov.–Dec. 1983, pp. 27–37.

Berry, Richard, ''Mars 5 Years After.'' Astronomy, July 1981, pp. 6–17.

Oberg, James E. (1982). Mission to Mars. Harrisburg, Pennsylvania: Stackpole Books.

Anon. ''Viking Mission.'' Astronomy, Oct. 1979, pp. 62, 63.

Wells, Ronald Allen. Geophysics of Mars. Elsevier Scientific: New York, 1979.

STUDY GUIDE:
Chapter Five Documentation

1. What is the purpose of documentation?
2. What is a reference note?
3. What is the most efficient documentation?
4. What is the general rule about how much documentation you should have in research writing?
5. Under what condition(s) might a student have very little documentation?
6. True or false: You cannot have too much documentation.
7. What is the string-of-pearls effect?
8. True or false: Paraphrases and summaries need footnotes.
9. When is a paper underresearched?
10. When is a researcher overworking the data?
11. When is a paper underdocumented?
12. What is plagiarism?
13. What is a content note?
14. When is a researcher showing off with content notes?
15. How can a content note be documented?

16. What are a few possible content notes?
17. True or false: If you change the words, or paraphrase, from a source and use only the idea, you need not document it.
18. What is the impact of plagiarism on credibility?
19. How may you document a major source?
20. If you want to use a quote (Smith) you find in a source (Brown), what are your options?
21. What are the objections to quoting from secondhand sources?
22. What is common knowledge?
23. True or false: Common knowledge need not be documented.
24. What is uncontested knowledge?
25. True or false: Paragraphs should not end with documentation.
26. True or false: APA style requires that you give the date along with authors' names.
27. What is the difference in the way MLA and APA treat multiple authors' names?
28. How should you give a citation note for an author who has more than one title in your bibliography?
29. What is the difference in the way MLA and APA identify page numbers?
30. What is the rule for inclusive page numbers, dates?
31. What is the difference in the way MLA and APA give authors' names in a bibliography?
32. What is the basic note form for a book?
33. How are items in a bibliography arranged?
34. How should subsequent books by one author be listed in MLA, APA, and traditional styles?
35. What is the basic note form for a periodical?
36. How can you give a note for an anonymous work?
37. True or false: You should not use a citation note to document a dictionary definition.
38. Why doesn't documentation for an encyclopedia source need page numbers?
39. What information is required when citing a newspaper article?
40. What is the difference in documentation between a magazine and a technical or specialty journal?
41. What information is required when citing a film?
42. If you collected information through an oral interview, how could you document it?
43. What does "pseud." stand for?

44. What are the standard abbreviations for May, June, and July? How do MLA and APA differ in treatment of months?
45. True or false: Writers avoid endnotes for Bible references.
46. Should you change roman numerals (volume numbers, for example) to arabic?
47. How do you write 1983 in roman numerals?
48. What number is given in roman numerals as DCLXVII?
49. What is the difference in the way APA and MLA treat volume numbers?
50. What is unusual about the way APA treats the first title in any References entry (for example, an article title)?
51. What is the difference between endnotes and footnotes?

CHAPTER SIX

Organizing Research Writing

The more data you collect, the more you willl need to find an effective strategy for organizing your research. When you have dozens of note cards to arrange into meaningful patterns, organization becomes your biggest problem. In general, examples and arguments should be arranged in order of importance, from least to most important, but you must come up with a specific plan for your paper.

DISCOVERING ORDER

As you collect your note cards, you can begin sorting them in different ways. At first you should be guided by the cards themselves; avoid imposing order on your data—try to discover the inherent order of your cards. Label your cards (lightly, in pencil) according to key ideas, main points. It's handy to have a dining room table or living room rug where you can spread out your cards and sort them into piles. Some researchers can do this on a computer, but to do so means having your notes on the computer in the first place. One alternative for computer users is to enter a very brief (one sentence) summary for each note card into a "Notes" file and then experiment with organizing these summary notes. At this point in the process you shouldn't concern yourself too much about the final plan for your paper. These early explorations should be tentative and experimental; look for whatever order may reveal itself in your notes.

Even if you happen upon a reasonable-looking plan at the outset, you should still experiment with other possibilities. A "good enough" plan isn't the goal. Look for the best plan. The notes on ape language, for example, at first will obviously fall into stacks of cards on each chimpanzee: Washoe notes,

Lana notes, Sarah notes, Nim notes, and so on. These stacks will begin to share certain similarities.

Washoe:	*Sarah:*
Learns a silent language (ASL)	Learns a silent language (tokens)
Can make sign for food, other requests	Can make sign for food, other requests
Invents new words, makes her own signs	Invents new words, makes her own signs
Can name trainers, other individuals	Can name trainers, other individuals
Can answer questions	Can answer questions
Can string words together, make sentences	Can string words together, make sentences

Thus, one simple plan would be to use each of the chimpanzees as an organizing principle. Since these ape experiments followed each other in more or less chronological order, one way to organize the paper would be to discuss each ape in turn. However, this plan contains a good deal of repetition; chimp after chimp learns a language, uses it to make requests, form sentences, and so on. There may be another way. Using the same note cards, we could stack them not by chimpanzee but by feature. Then we would get a strategy that looks like this:

Learns a silent language
  Washoe, Sarah, Lana, Lucy, Nim, Koko
Can make sign for food, other requests
  Washoe, Sarah, Lana, Lucy, Nim, Koko
Invents new words, makes own signs
  Washoe, Sarah, Lana, Lucy, Nim, Koko
Can name trainers, other individuals
  Washoe, Sarah, Lana, Lucy, Nim, Koko
Can answer questions
  Washoe, Sarah, Lana, Lucy, Nim, Koko
Can string words together, make sentences
  Washoe, Sarah, Lana, Lucy, Nim, Koko

This plan has a certain amount of appeal because, although it looks highly repetitive here, there are significant variations in each ape's mastery of each feature. Collecting all the apes under each feature might help to demonstrate their similarities and differences better than treating each ape separately. However, this way we lose the natural chronology. Within the features we could still discuss Washoe, Sarah, and Nim in sequence, but the features themselves don't quite occur that way.

You needn't be committed to an iron-clad decision about organization. As long as the reader can follow the development of your paper, you are free to use whatever plan best presents your material. You can decide, for example, to mix chronological and conceptual patterns, using time order for the overall strategy but occasionally switching to order of importance when necessary. You must keep prewriting, working with your notes until you find the organizational strategy that works best for the material you have and the audience you anticipate. The important thing is to have a plan. In your planning and revising cycles—before things settle into their final shape—keep experimenting, sorting, arranging your notes, and trying different organizational ideas.

THE FORMAL OUTLINE

After you have read your data and have come to know your subject thoroughly (you must assimilate your information), and have made your decisions about which organizational strategies to use, you will be ready to begin developing your outline. Most research papers require a formal outline. It's especially useful to anyone who reads your research, and it shows at a glance the structure and the points of development in your paper. Your instructor can tell from an outline whether you have your material clearly organized, and he or she can get an excellent preview of your work from a preliminary outline. If you have such an outline, your instructor (or a classmate) can offer advice and assistance before you write your paper.

As you collect your sources, begin to sort and organize them into major and minor points. This means, of course, that you must do some preliminary reading first, and your first outline should be considered only preliminary, subject to feedback. You may also make changes as you continue to collect data and think about what you are going to write. In other words, the outline isn't a simple, mechanical formula that you can invent and then simply fill in with details from your reading. For most researchers, the outline isn't so much a blueprint as it is an organic framework that grows and develops with the research. As you develop the outline, you begin to see the data in terms of categories you have set up. However, you mustn't permit your outline to become a mind set that you impose on the data. Instead of forcing data to fit into your outline, you must be prepared to alter the outline to accommodate the data. Even while you are writing the first draft of your paper, you may find yourself working back and forth, first shaping the paper to fit the outline and then altering the outline to fit the paper. This kind of creative give-and-take makes the outline a useful tool—it helps you understand what you are doing. When you finish, the paper and the outline will agree with each other. The logic of both will show your analysis of the data.

Revising the Preliminary Outline

Preliminary outlines are often pretty skimpy and sometimes not much more than a list. You must work back and forth between your notes, your developing paper, and your preliminary outline, revising for greater structure and detail as you go. A formal outline should go beyond listing the main points. The finished outline that accompanies a paper should be nearly as detailed as the paper. However, the preliminary outline submitted to an instructor needn't go beyond the third or fourth level of subdivision. At the outset, most research outlines look very similar:

I. Introduction
II. Body
III. Conclusion
IV. Works Cited

If you are reporting an experiment, a survey, or some other firsthand data, it's customary to describe the research procedures in a separate section:

I. Introduction
II. Review of Research
III. Methodology (description of procedures)
IV. Findings (presentation of data)
V. Conclusion
IV. References

Beyond the general structure of your paper, you must read your preliminary material to see what will go where in your outline. Read the following preliminary outline for a paper on the funeral industry.

Regulating the Funeral Industry

I. Introduction: Should the funeral industry be regulated?
II. Funeral industry must make a profit.
III. Funerals are expensive.
IV. Federal regulations would increase costs.
V. Consumers can shop for best prices.
VI. Conclusion: There is no need for regulation.

An outline like this doesn't have enough detail to tell anyone much about the paper. The writer has listed the main ideas, but there is no clue as to the structure of the paper or the relative importance of the ideas. Here all the ideas seem to be of equal importance. We can say two things of this outline: it doesn't seem that the writer has done enough reading yet, and it doesn't seem that there has been enough analysis of the data. The writer should ask what evidence there is for these statements. Why are they important? With additional work, the writer can produce a more helpful outline:

Regulating the Funeral Industry by Nancy Miller

I. Background to the Problem
  A. Today's funeral
    1. Quote
    2. Description
  B. FTC study and proposed rule
  C. Thesis question: Should the funeral industry be federally regulated?
II. Industry views against regulation
  A. The arguments
    1. Service industry—good reputation, consumer good will essential
    2. Federal regulation, increases in trend toward large chains
      a. Chains higher profit
      b. Chains more impersonal
    3. Reason for not advertising prices and not having unit prices
      a. Limits vital person-to-person contact
      b. Causes people not to choose some services valuable to them
      c. Could raise costs because of paperwork and advertising costs
      d. Could backfire into higher prices for each item
    4. House subcommittee—insufficient evidence to suspect abuse by funeral industry
    5. Proposed rules unfair to any business—federal regulations not preferred
    6. Funeral homes business establishments, seeking profit
    7. Violations of few generalized to entire industry
  B. Evaluation of the arguments
    1. Best argument—not apply regulations where none needed; not penalize entire industry for excesses of a few
    2. Emotional arguments—need for profit in funeral business
III. Arguments favoring regulation
  A. Opposing arguments
    1. Bargaining situation advantageous to funeral directors
    2. Methods of persuasion and pricing questionable
    3. Unnecessary embalming—excuses made for it
    4. Excess number of funeral homes

5. Leadership of funeral industry—effective organization
   a. Lobbies powerful
   b. State boards composed of funeral directors

B. Evaluation of arguments
   1. Opposing arguments valid
   2. Normal business procedures not in effect
      a. No competition
      b. Situations and methods favoring industry

IV. The need for regulation
   A. Insufficient regulation within industry at state levels
   B. Data against industry versus data favoring
   C. Need for federal legislation apparent

V. Works Cited

Anyone reading this outline can get a good view of the subject, the data, and the structure of the paper. There is enough detail to tell whether the writer has the paper well in hand.

Use Descriptive Labels

Since the Introduction and Conclusion are standard elements of any research paper, you should change these labels to more descriptive headings. Thus:

Regulating the Funeral Industry

I. Backgrounds to the question
II. Industry views of the question
III. Consumer views of the question
IV. The need for regulation

ACTIVITY 32 Prepare a preliminary outline for a report on your research topic. Let your outline cover only the key points, like the example above. Then review your research notes; reread two or three articles. Revise your preliminary outline with more detail. Try to get three or four levels of subdivision in your outline. Turn in both preliminary and revised outlines.

THE ABSTRACT

Not every paper requires an abstract. But the longer and more technical your paper is, the more an abstract will help your readers. If you are submitting a paper for publication, follow the guidelines of the journal to which you are submitting.

Obviously the abstract cannot be composed before your paper is written. Though the abstract appears first—either before or after your outline—it is written last. Read the following abstract, which accompanied a published paper by David Premack.

> ABSTRACT. Once the chimpanzee has been exposed to language training, it solves certain kinds of problems that it does not solve otherwise. Specifically, it can solve problems on a conceptual rather than a sensory basis. For example, it is only after it has been language trained that the normal chimpanzee can match, say, ¾ apple with ¾ cylinder of water, i.e., match equivalent proportions of objects that do not look alike. Similarly, it can match not only relations of sameness to sameness, but also difference to difference. The change from sensory to conceptual responding that language training appears to induce in the chimpanzee thus resembles a developmental transition that normal children undergo in the "natural" course of things—though whether the two changes are genuinely comparable is one of several ontogenetic questions that remain to be answered.
>
> David Premack, "Possible General Effects of Language Training on the Chimpanzee," *Human Development* 27 (1984): 268–81.

The APA guidelines suggest that most published abstracts should be not more than 100–150 words long. Much scientific writing calls for 250-word abstracts. *Dissertation Abstracts International* allows 600 words. See Appendix, Abstracts, for a model.

ACTIVITY 33 Find the guidelines on abstracts (in the front of the journal) in a professional journal in your field of interest. Assume you are an editor working with the author of the cocaine report in the Appendix; write an abstract to go with the article. Follow the guidelines for the journal. If you cannot find journal guidelines, write a 250-word abstract.

THE TITLE

Some writers compose the title first; others wait until the paper is finished before deciding on a title. Because research papers must be indexed so that other researchers can find them, your title is an important research tool. The title of a research paper should be informative, clear, and specific. A title like *The Old Scottish Dragon* is ambiguous and gives no clear guide to the subject of your paper. A better title might be *Is the Monster of Loch Ness Real?* Often the thesis statement or question is the title. Avoid cute or amusing titles.

Every paper must have a title, but the title is never considered part of your paper. Don't make any reference to the title or to the information in the title nor write your paper with the assumption that the reader knows the title.

For example:

> WHAT CAN BE DONE ABOUT THE DISEASES OF THE POOR?
>
> These are especially difficult to diagnose and treat, and the large numbers of such diseases make a social crisis for us.

The words "These" and "such diseases" make sense only if the reader refers back to "Diseases of the Poor" in the title. To avoid this connection between your text and your title, give the missing information in your introduction:

> WHAT CAN BE DONE ABOUT THE DISEASES OF THE POOR?
>
> The diseases of the poor are especially difficult to diagnose and treat, and the large numbers of such diseases make a social crisis for us.

Don't underline or put quotes around the title of your paper.

THE INTRODUCTION

In some scientific writing (reports not meant for publication) researchers may dispense with the formal introduction. However, published research needs an introduction for the same reason that everything else does—to catch reader interest. In the introduction you must make your thesis statement or ask your thesis question. Whether to use a question or a statement is the writer's choice. The thesis question is useful in some situations (check with your instructor) because a question doesn't give away at the outset what conclusions your paper will reach. On the other hand, the thesis statement announces from the beginning what your paper will illustrate and serves as a guide for some readers. Some instructors prefer to see the thesis as the first sentence of the introduction; however, it also makes good sense to have it as the last sentence. If your thesis finishes the introduction, it provides a good transition to the rest of the paper.

- An introduction can be a single paragraph or longer, if it doesn't delay the paper unnecessarily. The length of the introduction depends somewhat on your purpose. If you are writing a historical background or discussing the significance of your research question, the introduction may be a separate section of your paper, several paragraphs long.
- Avoid one-sentence introductions. Remove such sentences entirely, connect them to the next paragraph, or build them into more substantial paragraphs.
- Also avoid empty introductions that wander vaguely around the subject. The introduction isn't an ornament: it's an important part of the paper and should provide the reader with useful information.

- Your introduction shouldn't assume the reader knows the title to the paper, the assignment, or anything else. Even though you may be writing the paper at the direction of your instructor, the paper must stand on its own. Make no references to "the assignment" or anything else not in the research. You may wish to use one of the standard introductory strategies. (See p. 214.)

Your introduction, like the paper itself, depends on your situation: What are you writing about, who are you writing to, what is your purpose? Read the two introductory excerpts below. How can you account for the differences in them?

> Comparisons we have made in recent years between chimpanzees that have and have not been language trained suggest that the training may confer some general advantages. The suggested change is at least reminiscent of the kind of transition one sees in a child as it moves from a sensory stage of processing to a more conceptual one. To be more specific, we find that language training appears to give the animal an advantage for a particular class or kind of task; on another kind of task—one we have come to call "natural reasoning"—we find no difference between animals that have and have not been language trained. One of our aims will be to distinguish between these two kinds of tasks. Of course, ultimately we wish to ask the *ontogenetic* question, what we might call *Piaget's* question, viz. What is the nature of this transition, this change that we seek to characterize by contrasting "sensory" with "conceptual" (neither of which are entirely definable terms)? Is the change that the language training appears to bring about in the chimpanzee truly like the change we observe in the child? In addition, we would like to address the *phylogenetic* question: If a species more specialized in intelligence than another brings special training to bear upon the latter, why should the latter undergo a change of the kind that we appear to see? Before addressing these two questions, let me first give a brief characterization of the language training—not blow by blow detail, not who bites whom and how many times—but the structure of the training.
>
> David Premack, "Possible General Effects." *Human Development* 27 (1984): 268–81.

Premack's paper is a description of the kind of language training the chimps had. The point of his paper is that there are striking differences between these chimps and others not so trained.

> Sarah hangs by one long arm from the side of her cage and stares out at her visitors. Her lambent brown eyes consider us gravely. Meeting her this way, finally, after reading so much about her linguistic accomplishments, is in an odd way unsettling because there is nothing to say and it seems that there should be. It feels awkward to simply stand there meeting her gaze, returning stare for wordless stare, in almost the same way it would feel socially awkward and frustrating to meet a writer whose books you've read in translation only to find that

> you share no common language and cannot make conversation, can't exchange so much as a hello to civilize your exchange of stares.
>
> Sarah, betraying no feeling of awkwardness, watches us meditatively, dabbing with one finger at her runny nose. Her teacher, Amy Samuels, offers her a paper towel: "Here, Sarah—blow your nose?" Sarah accepts the paper towel, takes a swipe at her nose, and lets the paper fall to the wet floor of her cage. She apparently finds us more interesting and continues to watch.
>
> As Bruce Stromberg, who photographed this story, said, "It gives you a weird feeling—you know she's thinking about you, sizing you up, deciding if she likes you or not, but you can't tell *what* she's thinking."
>
> Patricia McLaughlin, "Learning from Some Chimps." *Pennsylvania Gazette*, February, 1976: 16–22.

Despite this introduction, McLaughlin's article is not about her visit with Sarah. Instead the article is a summary of the ape-language research, Professor Premack's involvement with this line of research, and a report of Sarah's language-training program. What the writer, McLaughlin, saw the day she went to see Sarah is used here as an introductory strategy to get the reader interested in a review of research.

ACTIVITY 34 Write your analysis of these two introductions. How are they different? Cite examples of the differences. Give your analysis of the cause of the differences: Why are the writers writing this way?

ACTIVITY 35 Assume you have been invited to submit your autobiography to the journal *Human Development*, and also to the *Pennsylvania Gazette*. Analyze your purpose and point of view for such an assignment: What voice should you project? What tone should you use with the subject? What attitude should you express toward your audience? Write the introduction only, not more than a page.

Introductory Strategies

Start with a dramatic incident you find in the research, such as the Mount St. Helens explosion, in a paper on volcanoes.

Start with a contrast, such as today's four-cylinder car versus the great gas guzzlers, in a paper on oil prices.

Start with a story relevant to the subject, such as the incident of General Patton's slapping a soldier, in a paper on the modern army.

Start by setting the scene, such as a description of California's beaches prior to an oil spill, in a paper on pollution.

Start with a question or problem, such as the thesis question or some related question (Have you planned for the cost of your funeral?) in a paper on the funeral industry.

Start with a description, such as the appearance of the refugees in resettlement camps, in a paper on foreign aid.

Start by explaining the thesis, such as the meaning of the Equal Rights Amendment, in a paper on the ERA.

Start with a historical review, such as our history with prohibition, in a paper on legalizing marijuana.

Start with unusual facts and figures, such as the amount of the national deficit, in a paper on the cost of welfare.

Start with a quotation, such as a relevant quote you find in your research or an older, perhaps memorable quote relevant to the subject—a quote you may find in a book of quotations.

Start with a definition, such as the definition of the key term or some significant concept in the research, in a paper on euthanasia.

Start with an idea to be refuted, such as a misconception, myth, or stereotype, in a paper on mental retardation.

THE BODY OF THE PAPER

Thesis and Support

The primary organizational strategy of much writing is thesis and support. Your introduction states a thesis, such as "New drugs hold a promise for AIDS victims," and the rest of your paper will cite research that demonstrates (illustrates) this thesis. Alternatively, you can form your thesis in the shape of a question: "Do new drugs hold a promise for AIDS victims?" The rest of your paper will then cite evidence answering the question. (See Chapter One.)

General to Specific

By far the most popular organization of the thesis and its supportive statements is deductive. The paper progresses from the general statement (the thesis) to the specific supportive statements. Whether you are organizing the entire paper, sections of it, or even paragraphs, most readers expect to encounter the general statement first (your thesis), followed by your specific evidence. If you use the general-to-specific order, your paper will have a conventional structure. The two basic kinds of support are reasons and examples:

Argument (Opinion) and Reasons

[opinion]
The weeping willow is a fine tree.
[reasons]
[because] It grows quickly, becoming a fine shade tree in just a few years.

[because] It is a hardy tree, resisting insects, blight, and drought.
[because] It is a very attractive tree; its long, "weeping" branches are very graceful and romantic.
[because] The supple willow branches can be curved, bent, and shaped into attractive and useful baskets and furniture.
[because] It will grow in wet, swampy areas where other trees won't.

Assertion and Examples

[assertion]
The language-learning apes have passed every language test.
[examples]
[for example] They have used language to make and respond to requests.
[for example] Chimps have matched objects and pictures of objects with names; they can "label" with language.
[for example] They have invented new words; they are creative with language.
[for example] They have strung words together to form sentences; they can learn grammar.
[for example] They have discussed objects not present; they can answer questions about time.
[for example] They have sorted and classified words.
[for example] Apes have told lies and made jokes.

ACTIVITY 36 Brainstorm for events or activities you have experienced or read about that you would call either pleasant or unpleasant, positive or negative. Write a thesis statement. For example: "Having a car on campus is a nuisance." Then list as many reasons or examples as you can think of to illustrate general-to-specific arrangement.

Specific to General

An equally valid arrangement of thesis and support, "inductive" order is the reverse of general to specific: the paper develops from the specific supportive statements to the general conclusion—the thesis. Some writers prefer this arrangement since it produces the effect of reasoning with the reader; it can look almost syllogistic. That is, the paper seems to be saying, Premise A is true, premise B is true, premise C is true; therefore, we must reach conclusion D.

[since] They have used language to make and respond to requests.
[since] Chimps have matched objects and pictures of objects with names; they can "label" with language.
[since] They have invented new words; they are creative with language.

[since] They have strung words together to form sentences; they can learn grammar.
[since] They have discussed objects not present; they can answer questions about time.
[since] They can have sorted and classified words.
[since] Apes have told lies, made jokes.
[therefore] The language-learning apes have passed every language test.

ACTIVITY 37 Brainstorm your research topic or your own experiences for a thesis you can support with reasons or examples. For example: "The library is a sophisticated information retrieval system." Then list your reasons and examples first; end with the thesis to illustrate specific-to-general organization.

For most research writing, the simplest organizational patterns are usually the best.

Chronological

Often the simplest plan is chronological order; start at the beginning and proceed in order of events to the end.

> A group of five objects was placed on a tray outside one of the test rooms by one of the experimenters. While the chimpanzee was inside the room, it could not see any of these objects. Out of view of a second experimenter, the chimpanzee came to the door, looked at the five objects in the tray (see figure 8), walked back into the room, and used the keyboard to indicate which of the five objects it was going to give to the second experimenter (see Figure 9). The chimpanzee then returned to the tray of objects, picked up one of them (Figure 9), and gave it to the second experimenter (Figure 10), who could not see the chimpanzee while it surveyed the objects on the tray, while it lighted the symbol on the keyboard, or while it picked up the object it was going to give to the second experimenter. This experimenter could, however, determine from the projectors outside the room which item the chimpanzee said it would give. Once the chimpanzee came around the corner to give the object, if the object and the lighted symbol were the same, the second experimenter praised and rewarded the chimpanzee. If not, the teacher expressed dismay and puzzlement and proceeded with the next trial. On each new trial, a new set of five randomly determined objects was placed on the tray by the first experimenter.
>
> Sue Savage-Rumbaugh, et al., "Can a Chimpanzee Make a Statement?" *Journal of Experimental Psychology: General,* 112 (1983): 481.

ACTIVITY 38 Prewrite ideas for a paragraph developed by time order. You may use research topics or firsthand data. Assume you are writing for the *Journal of Experimental Psychology*. Write one paragraph (not more than a page).

Spatial

If you are writing a description, you should start at some reasonable point and then move in orderly fashion from one point to the next. A description of an automobile, for example, probably should not start with the carburetor. If you are describing the external features of a car, it does not make much sense to hop from the headlights to the tailpipe. There is no prescribed order in a description, but whatever order you do follow should have some logic the reader can see and understand. Read the following descriptive excerpt. How is it organized?

> The metal shed that houses the siamangs, squirrel monkeys, and mandrill also houses the training area for the instruction and testing of Ameslan. It contains a raised platform from which observers can record the chimps' signing behavior. In more natural situations, like the chimp island, Fouts uses videotapes as well as observers to record the chimps' behavior.
>
> About 100 yards east of the main yard is the small lake that contains the three man-made islands. On these islands roam three colonies of primates restrained only by the water isolating little tracts of land, which more naturally serves the function of iron bars. One island is grassy and surmounted by two tall poles connected by a rope, which is there for the benefit of a small colony of new world or capuchin monkeys. Next in this archipelago is another islet of about a half acre. In contrast, it is densely foliated with cottonwoods and willows, which provide an arboreal habitat for the gibbons. Of all primates, the gibbon is perhaps the most perfectly adapted to life in the trees. It can flash through thick branches faster than a man can run and has the equilibrium, timing, and confidence that would shame any mortal high-wire performer.
>
> As noted, the gibbons' entertainment is provided by the colony of juvenile chimpanzees who inhabit the third island, on which, instead of trees, there are tall poles. For shelter there is a brown African hut, a "rundevaal," which is heated in the winter and is equipped for videotape observations of its occupants. The chimps on the island are all juveniles, and all have had training in Ameslan. Thelma, Booee, Bruno, and Washoe spend most of their time here, either sporting about the hut or sitting contemplatively atop the poles.
>
> Eugene Linden, *Apes, Men, and Language*, p. 87

Another kind of description often found in research writing is more technical, but shares the same general purpose of explaining by describing. For example:

> **PULLEY AND BELT,** a mechanical arrangement for transmitting torque from one shaft to another. The torque originates at a wheel or drum, called the drive pulley, that applies torque to a continuous, flexible member called the belt. The belt passes around and delivers torque to a second wheel or drum, called the driven pulley. A pulley may be made fast to a shaft by a setscrew or key in the pulley hub, or by a tapered bushing drawn between shaft and hub; or the pulley may turn freely on a shaft as an idler. Pulley rim and hub may be one solid piece

connected by a flange, web or tapered arms, depending on the pulley diameter. Pulleys whose peripheral surfaces have circumferential grooves are called sheaves.

Excerpt from "Pulley and Belt," *Encyclopaedia Britannica*, 1970

ACTIVITY 39 Compare the two descriptive excerpts above. How are they different? Why are they different? Write your analysis of purpose and point of view (voice, tone, attitude) in the two excerpts.

Order of Importance

When you have more than one argument, more than one example, more than one detail, which one comes first? Which comes last? The most common answer to this is to give the least important first and the most important last. That is, the structure of the paper ought to be based on order of importance—from least to most important. This is counterintuitive for most students, many of whom prefer to give their most important information first, perhaps due to the journalistic principle which allows editors to cut from the end of an article to make it fit the available space. However, a research paper that becomes increasingly less important as it progresses will have the wrong effect on readers.

Still, the order of importance need not be applied mechanically. There should be enough flexibility in your structure to allow some variation, even within the overriding strategy. If you have more than one important point—for example, two important points and two or three lesser points—you may choose to use the "covered wagon" strategy: place the weaker points in the middle, sandwiched between the strong points. Overall it is more important that you achieve the effect you want with your readers than that you follow rules and patterns of organization. Look at your outline; ask yourself why your points are arranged in this order. What effect are you trying to achieve?

Making Concessions

You are required to concede whatever must be conceded. You cannot bias your research by reporting only the data you approve of. You are not supposed to attempt to prove anything in research writing. Even in an argumentative paper, you will show yourself to be reasonable and unbiased if you concede points to the opposition. Besides, when a point must be conceded (i.e., it is undeniable) you lose nothing by making the concession—and you gain credibility.

When you write about an issue or controversial question with two sides (or more) to investigate, you must present yourself as a disinterested researcher. (See Impartiality, p. 117.) You must deal with a problem of bias involved in the order in which you present information. The best choice is to present the weaker view first (develop by order of importance). If your research has convinced you that cigarette smoking causes cancer, you must first present the other side, the side of those who feel that cigarette smoking doesn't cause cancer. Presenting the stronger side first doesn't work; the weaker side then seems irrelevant merely as a consequence of following the arguments you have already established as stronger. Presenting the weaker side first, and presenting it well, will help to convince the reader that you understand the issues, and will show that you are a fair and unbiased researcher. You must present the reasons and the evidence that the opposition would use; make the case for the opposition as strong as it can be made. As an impartial researcher, it is your duty to act as advocate for both sides of any issue. You are not the prosecutor but the judge, evaluating all evidence impartially. Presenting the weaker side first ensures that that side will get a fair hearing. Putting your strongest argument in the strongest position—last—will give your paper a natural progression ending with strength.

Don't offer counterevidence here; don't find flaws in the opposing view. It's poor strategy to introduce a point for the purpose of refuting it. For example: "Nonsmokers complain about the smell of cigarettes in restaurants. However, this isn't a good argument. They can easily move to the nonsmoking section every restaurant has today." Here the researcher has introduced the argument about the smell of cigarette smoke, a legitimate argument against smoking in public. Then the researcher makes the mistake of immediately countering the argument with a refutation. This strategy makes the researcher seem argumentative. It sounds like the researcher is manipulating the data in order to win the argument, deliberately introducing an argument so that it can be countered. Thus, there should be no arguing with the data here. The point of this section is to present the opposing side as well as possible. You must concede as much of the opposing view as the facts require. It is possible to concede all the opposing evidence yet still show that the other side has more data, better research, and so on. Leave faultfinding until you begin to evaluate the evidence. Remember the rule of impartiality (Chapter 4): Researchers are not supposed to "prove" or "win" anything. Your job is to find the truth, whatever it is.

ORGANIZING PARAGRAPHS

Too much bad writing can be traced to authors who merely "grow" their paragraphs instead of deliberately planning them. In the drafting stage of your work you may "grow" paragraphs, if that helps you to generate your first draft. But in the revision stage you should be hypercritical of such paragraphs. A paragraph, by definition, should discuss one and only one idea. Both you and your reader should be able to see that idea clearly in each of your paragraphs. Ambiguous, loosely written, poorly developed paragraphs make hash of research. For example, read the following:

> Washoe was the first of the "scientific" chimpanzees. Her trainers kept records of her language behavior. Washoe used ASL to make signs for things she wanted. She was raised like a child in the Gardners' home. Washoe learned to make the sign for "out" when she wanted to go outside. The world was amazed by the films of Washoe signing. Washoe was the first real "talking" chimpanzee.

Each of your paragraphs must submit to the paragraph test. You must be able to answer the questions, What is this paragraph about? Why is it here? What function does it serve? In a general way you could say this paragraph is "about" Washoe, but that's an insufficient answer. *What* does it discuss about Washoe? Some of the sentences are about Washoe's use of sign language—but not all of them. This is an example of a paragraph that has "grown" all by itself (more or less), in a very loose fashion. The writer has permitted him- or herself to daydream through the paragraph, allowing free association to be the organizing principle. In revision you must rewrite paragraphs like this. If you decide that you want to write about Washoe's use of sign language, you should get rid of everything else, rethink the point you are trying to make, and then deliberately restructure so both you and your reader will be able to see the logic of your paragraph.

Paragraph Strategies

Use an Organizational Plan

Professional writers usually don't resort to formulas for paragraph writing; they rely on their "sense" of what goes with what and their experience with research writing. However, student writers will find they can begin to develop this kind of "paragraph sense" if they do keep in mind a general formula for research paragraphs: assertion, expansion, substantiation, conclusion.

- *Assertion* The writer begins with an assertion, a statement of the point the paragraph will make. See "topic sentence" below.
- *Expansion* The writer expands on the assertion, clarifies it, adds to it, makes it more specific.
- *Substantiation* The writer offers support for the expanded assertion: examples, arguments, quotations, comparisons and analogies, and the like. The more supportive material, the better.
- *Conclusion* The paragraph may or may not end with a final comment, such as a reference back to the topic statement.

Admittedly not many professionals' paragraphs follow just this formula. There are too many ways to vary paragraphs, but you can see from this a general plan you can use to give order and development to your own paragraphs. (See paragraphs by Kidder, p. 223, and Attenborough, p. 225.)

Use a Topic Sentence

The most obvious strategy for making sure that a paragraph is unified (discusses a single idea) and that your reader will recognize the unifying idea is to state that idea explicitly in a *topic sentence*: "Washoe uses American Sign Language to communicate with her trainers." Any time (but especially during revision) there is any doubt in your own mind about the unity and clarity of one of your paragraphs, look for its topic sentence, the sentence that says what the paragraph is about. If there is no topic sentence, supply one. If you cannot think of a single sentence that will incorporate all the ideas in your paragraph, revise the paragraph. The topic sentence can appear first, last, or in the middle of your paragraph. Its position is less important than the ease with which it fits into the paragraph—the logic of its position relative to the other sentences. Read the following paragraph. Is it clear to you that the first sentence is the topic sentence?

> *One laser printer may offer some convenience features that another doesn't.* For example, Okidata's Laserline 6 weighs in at 33 pounds and has a warm-up time of just 35 seconds. Genicom's 5010 is nearly silent, but weighs a massive 100 pounds and is ready to work 95 seconds after being turned on.
>
> Heather-Jo Taferner, "Laser Printers," *Personal Computing*, Aug. 1987: 125.

ACTIVITY 40 Prewrite a paragraph about computers, typewriters, cameras, or other technology you are familiar with. Revise your paragraph until it contains a topic sentence that clearly identifies the subject of your paragraph.

Develop with Information

The development of a paragraph refers to how many details or how much information you have and how it is organized. A well-developed paragraph has ample data, illustrates a single point, and is clearly and efficiently organized.

1

Other kinds of ants tap the nutriment of grass by using as an intermediary, not fungi, but aphids. These insects digest only a small part of the sap they suck. The rest they excrete as a sugary liquid that is known, somewhat flatteringly, as honey-dew. It can often be found as a sticky film on the ground beneath an aphid-infested plant in the garden. Some ants, however, find honey-dew an excellent food and they herd the aphids into flocks and milk them in much the same way as human farmers tend their herds of dairy cattle. The ants encourage the aphids to produce more honey-dew than they would normally do by stroking them repeatedly with their antennae. They protect them by driving away any other insect invading the aphids' grazing grounds with squirted volleys of formic acid. Some build special shelters of parchment or earth around a particularly productive stem, or root on which the aphids are grazing and so deprive them of their free-range, like animals in a factory farm. At the end of the summer, when the aphids die, the ant-farmers take the aphid eggs down to their nests for safe-keeping. When young aphids hatch out in the following spring, the ants carry them out again and put them to graze on fresh pastures.

Topic statement

Expansion

Six supportive items

David Attenborough, The Living Planet, p. 118

Note the number of things the author mentions in each sentence. The "information load" of his sentences is heavy. It is the details, after all, that distinguish research writing from other kinds of writing. The number and quality of details, the ease with which you can discuss details, the skill with which you can incorporate details into your sentences—these make the difference between a merely competent researcher and a superior research writer. Developing by detail is the chief strategy of much research writing.

ACTIVITY 41 Prewrite the subject you feel most expert in. List or map out as many details as you can. Write a single paragraph developed by detail. The paragraph is a test for applicants applying for the job of editor of a professional journal.

Use Unifying Devices

Often the writer is the last to know that a paragraph (or entire composition) is not unified. Occasionally the reader will not know exactly what is wrong with a paragraph except that it isn't clear or doesn't stick to the subject. To prevent confusion and to make your paragraphs effective, use unifying and signaling devices like transitional signals, pronouns, synonyms, and repetitions of concepts. Read the following paragraph. Note the many different ways the author has tied his ideas together with unifying devices.

> When **Goma** was nearly {a year old}, the Langs obtained a {young} male (gorilla) of the same age for **her** as a playmate. [This], of course, was better than any number of human {children} for the normal development of the (gorilla). At first **Goma** was plainly jealous, determined to retain **her** superiority, and **she** tried to beat **him** into submission, but by rather cowardly methods: "**She** remained haughtily at a distance, **her** hair still standing on end. Then **she** reached out quickly—as if by chance—to **her** fellow creature, but immediately made for safety behind me, climbed onto my knee again, and into my arms. Some time later **she** tried to give **him** a quick bite or pluck at **his** hair. [This] situation continued for a long time. Often **she** bit **Pepe's** skin quite deliberately, but **he** was not in the least timid, and bit back." **Pepe** knew how to deal with **Goma**, and ten days later the two {young} (gorillas) had sorted it out, and were playing in a friendly way. **Goma** even allowed **Pepe** to bite **her** now, in fun of course, but one feels that **Pepe**, even at {a year old}, had asserted **his** masculine dominance.
>
> Vernon Reynolds, *The Apes*, p. 172

The author, Reynolds, has used many devices to tie this paragraph together so that the reader cannot get lost or become confused about the subject. The primary device, in bold print, is the chain of pronouns and proper names identifying the subject, the two apes. Time signals and references to time, underlined, help to keep clear when things happen, and they form bridges from one sentence to the next. A key term like "gorilla" (in parentheses) is repeated and thus echoes back to its earlier use. In two places Reynolds has used demonstrative pronouns [in square brackets] to point back to preceding sentences, and a string of synonyms {in braces} resonates the idea of children throughout the paragraph.

ACTIVITY 42 Write a paragraph of your own, using unifying devices. Revise your paragrapn until it is as thoroughly unified as the Reynolds's paragraph.

THE CONCLUSION

The conclusion of a research paper shouldn't be a mere summary; it should be the climax of the paper. There is important work to do in the conclusion. For example:

1. Review the key points for your readers.
2. Evaluate the key points.
3. Answer your research question.

So far, your paper has shown there is a research question and evidence supporting it and/or contradicting it. Now you must evaluate the data for your readers. Are all the arguments equally valid? Are all the data equally good? To evaluate the data you must point out for your reader which are the strong arguments and which are not so strong. And now you must *answer your thesis question.* It's a good idea to start your answer with a brief review of the arguments. The purpose of this mini-summary is to remind the reader of the basic positions in the research.

Students tend to write conclusions that are too short. You must remember that the reader hasn't read the material in the library, hasn't spent hours analyzing the data. The conclusion may seem obvious to you, but it's your job to help the reader understand what you understand. You must conclude one of two things: either the evidence leads to a conclusion, or there isn't enough evidence available for any intelligent conclusion to be drawn. You must help the reader understand the conclusion. Do you need better data? More data? Newer data? Better authorities? Fewer holes in arguments? More convincing or more logical arguments? Perhaps the whole question will hinge on some key point (the major underlying assumption is wrong, for example).

You should save something good for the conclusion. The conclusion isn't just the end of the paper; it's the *point* of the paper. A researcher is a detective. After examining all the evidence, you must solve the mystery, the puzzle. Everything in the paper must aim at the conclusion, where you will tell the reader what the research means. Researchers in a hurry often read only the conclusion of a research report to see whether it's relevant to their work. To such researchers the conclusion is the most important part of the paper (why waste time reading a long report only to find in the conclusion that there was no significant finding?). Reading the conclusion first is a research shortcut. Your conclusion mustn't be dismissed with a short concluding paragraph. If you use up all your good material in the body of the paper, your conclusion will look weak by comparison. Think of the conclusion as a short essay that could be read independently of the rest of the paper. In the rest of the paper, you present data; in the conclusion you reason about the data, showing the reader how an educated person *should* reason about the data.

In addition to showing the reader the outcome of the argument, you must bring the paper to completion and give the reader a sense of ending. A good quote, some striking fact or statistic, a relevant personal note, or just an especially well-worded final sentence will round off the conclusion satisfactorily. Frequently research papers end with a call for more research, an investigation of questions uncovered by the present study. Avoid raising any new or irrelevant questions in the conclusion, however.

Avoid Increasing the Charge

If you have demonstrated that cigarettes have been linked to cancer, emphysema, and heart attacks, you must resist the temptation to increase the charge ("and perhaps they will one day be linked to arthritis and insanity as well").

Avoid Extending by Analogy

You mustn't attempt to extend the argument by easy analogy ("and if cigarettes are so bad, we must suppose that cigars, pipes, and chewing tobacco are worse"). No matter how reasonable such an extension seems, it isn't supported by the data in your research, which concerned only cigarettes. To "suppose" that there is, or even "may be" such a relationship amounts to guessing—not research. You may however, propose that future research should investigate whether there may be a relationship between cancer, and other diseases, and other forms of tobacco use, such as cigars, pipes, and so on.

REFERENCES

Different instructors may require different formats for documentation. Before handing in documented papers, find out which format is preferred. Footnotes and endnotes are no longer used in many disciplines. Preferred modern practice is to give author's name and other information in your paper, like this (Smith 295). At the end of the paper, your reader will find Smith's work listed in your bibliography. The two most widely used styles for in-text references are the in-text style used by the Modern Language Association and the name-

and-date style used by the American Psychological Association. Both are illustrated in Chapter Five. If there are very few references, you may give the full bibliographic information in your paper itself. For example:

> According to Kyle Blane in his article "Research Problems," *Research Quarterly*, 15 No. 2 (1981), 22–25, this area calls for much work.

BIBLIOGRAPHY

There are different kinds of bibliographies, each serving a different purpose. A bibliography can be a useful part of your paper; the bibliography lets your reader see how recent your research is, and shows whether you are aware of important publications. Thus, it's a measure of how thorough your research is, and it becomes an aid to other researchers.

You will undoubtedly read more than the works you cite. Ideally you should read and assimilate everything available on a research question, but practically, there is likely to be too much material for such thoroughness in anything shorter than a doctoral dissertation. Nevertheless, you can skim through a great number of sources to learn how relevant they are. You will soon get to know from a little reading which are the important books, which things you must read. Read as much as possible of the available material in the time you have. Never put anything into your bibliography that you haven't actually seen; you must not simply copy a bibliography from the indexes. Skim through each item to see that it's in fact related and likely to be useful.

Follow the directions of your instructor. Use the kind of bibliography that serves your purpose: An *annotated bibliography* gives your descriptions or evaluations of the sources. The *limited bibliography* is limited to just those works actually cited in your paper (that is, the bibliography matches your in-text references). The *selected bibliography* includes everything you read for your paper, regardless of whether you actually cited all of it in the paper. Both the new MLA style and the APA style always require the limited bibliography only.

Alphabetical Order

Items in a bibliography are listed alphabetically according to the last name of the author or the first word of the title when no author is given.

Alphabetize word by word:

> "Men at Work."
> *Menace at Home.*
> "Menagerie."
> Mencken, H. L.

Authors' names starting with an article (*a, an, the*) should be alphabetized with the article moved to the end:

American Nuclear Society, The.

In formal writing, authorless titles beginning with an article are alphabetized as if they started with the next word in the title. Do not alter titles: "The Asteroids Are Coming" would retain its article at the beginning but would be alphabetized under "Asteroids"; "The Fancy New Video Games" would be alphabetized under "Fancy."

Authorless titles starting with numerals, signs, or other nonword symbols should be alphabetized as if spelled out: "1989, The Year of the Chimpanzee" is listed as if written "Nineteen Eighty-Nine."

ACTIVITY 43 Read the following rough draft of a research paper carefully. Review Chapters One through Six and then write your analysis of the paper's title, introduction, body, conclusion, and references. Assume you are the editor for a business journal. What are the paper's strengths and weaknesses? If there are weaknesses, can they be revised? What advice would you give the author?

CORRECTIVE ADVERTISING

Corrective advertising is to correct and retract misleading advertising with a solid dose of accurate information. As proposed by the Federal Trade Commission, corrective advertising is not only beneficial to the consumer, but necessary. The consumer lacks the knowledge to make intelligent choices of products. Mass production, rapidly increasing technology, and product differentiation makes the consumers job increasingly difficult. No longer is there only one kind of laundry soap etc. to choose from. ''Technological change is so rapid that the customer who bothers to learn about a commodity or a service soon finds his knowledge obsolete. In addition many improvements in quality and performance are below the threshold of perception and imaginative marketing often makes rational choice even more of a problem.''[1] Not able to distinguish the better product, the consumer can judge products only on the basis of the information he receives in advertisements. The consumer then associates quality with brand names. A perfect selling market is thus created in which the seller's product is not graded on its own merits, but on the invented merits claimed by the seller in his advertising. An example of such an ad, which was ordered to be corrected by the FTC, was one created by Ocean Spray cranberry juice in which they claim ''Ocean Spray Cranberry Juice Cocktail has more food energy than orange or tomato juice.'' The ad does not define ''food energy'' nor substantiate the claim that Ocean has more of it. The following is the corrected ad. ''If you've wondered what some of our earlier advertising meant when we said Ocean Spray Cranberry Juice Cocktail has more food energy than orange juice or tomato juice, let us make it clear: we didn't mean vitamins and minerals. Food energy means calories. Nothing more. Food energy is important at breakfast since many of us may not get enough calories, or food energy, to get off to a good start. Ocean Spray Cranberry Juice Cocktail helps because it contains more food energy than most other breakfast drinks.''[2]

In response to accusations of deceptive advertising, business claims that ads are not intended to be believed. This ''pseudo-truth'' is defined as ''a false statement made as if it were true but not intended to be believed. No proof is offered for pseudo-truth.''[3]

The FTC evaluates ads for the consumer on the basis of the ''economic man'' theory. That is, the FTC views the consumer as an intelligent, reasoning being who utilizes objective values to maximize his dollar. The key ingredient for the consumer then, is accurate information. Business, however, disagrees with this theory of the consumer. The consumer is viewed as an emotional being whose buying patterns are controlled by attitudes and habits. Advertisements today play on the consumers' emotions. Both of these viewpoints are partially correct, as man is an emotional being. Accurate information and emotional appeal could and should be placed in the same ad. The FTC is simply saying that more information should be contained in advertising so the consumer can make a rational choice. ''The integrity of the whole marketing system depends in no small part upon adequate buyer information.''[4]

Public disclosure of accurate information would also enhance competition. Companies would be forced to challenge each other's claims dealing with safety, performance and comparative price. Companies with evidence to substantiate their claims would be able to use the government, a people approved channel, to sell their product. Pitofsky, head of the FTC, feels that consumers will appreciate the candor of advertisers who they've been misleading in the past.[5] Business is afraid that disclosure of such information will lead to product suicide. An interesting example is that of Profile bread. The FTC found unsubstantiated claims and ordered corrective advertisements in the amount of 25% of Profile's advertising budget. The consumer response was so good that Profile raised the amount of corrective advertisements to 60%.[6]

Individual companies have mixed emotions about corrective advertising. Marywell Lawrence, Chairman of Wells, Rich, Green stated, ''I am delighted about it. It will raise our credibility with the public.''[7] Corrective Advertising would also help remove nondistinctive products like aspirin or bleach. With no real difference between them, these products would conceivably be the only ones hurt by corrective advertising.

A major area that needs correcting is advertising directed towards children. The FTC has already banned ads featuring premiums like little toys and trinkets, and is now investigating ads about dangerous toys and children's vitamins. Broadcasters, however, feel any restrictions on children's advertising would lower the quality of children's programs since sponsors, in paying for their ads, keep the shows on the air. Ads for vitamins, directed towards children have caused the incidence of vitamin poisoning to increase, and it is now the second largest cause of poisoning in children under 5.[8] Vitamins like Pals, Chocks, and Flintstones do not even have the Child Safe Caps, and an overdose can be fatal.

> Erin Shelton, now age five, said that Captain Kangaroo had told him that vitamins make a little boy grow big and strong. He wanted to grow big and strong real fast. That's why Erin ate 40 Pals vitamins with iron.[9]

Corrective advertising, necessary to the consumer, can also benefit business that have a quality product. Non–distinctive and inferior quality products will be shown as such, making it easier for the consumer to choose the better product. Children's vitamins won't be advertised as candy, although children's programs may suffer. This is a small price to pay for a decrease in vitamin poisoning. Corrective advertising must be approved, there is no real substitute for the truth.

4

End Notes

[1] Jack L. Taylor, The Consumer in American Society: Additional Dimensions (New York: McGraw Hill., 1974), p. 66.

[2] Arch W. Troelstrup, The Consumer in American Society (New York: McGraw Hill, Inc., 1974), p. 48.

[3] Ibid. p. 49.

[4] Arch W. Troelstrup, The Consumer in American Society (New York: McGraw-Hill, Inc., 1974), p. 52.

[5] ''Ally for Admen against FTC,'' Business Week, June 19, 1971, p. 39.

[6] ''Admen burn over baker's bow to FTC,'' Business Week, July 10, 1971, p. 25-6.

[7] Time Magazine June 14, 1971, p. 82.

[8] Jack L. Taylor, The Consumer in American Society: Additional Dimensions (New York: McGraw-Hill, Inc., 1974), p. 162.

[9] Jack L. Taylor, The Consumer in American Society: Additional Dimensions (New York: McGraw-Hill, Inc., 1974), p. 162.

Bibliography

''Admen burn over baker's bow to FTC,'' Business Week, 10 July 1971, p. 25-26.

''Advertising Claims to get Closer Scrutiny,'' U.S. News and World Report, 21 June 1971, p. 53.

''Ally for Admen against FTC,'' Business Week, 19 June 1971, p. 39.

''Burden of Proof,'' Newsweek Magazine, 21 June 1971, p. 22.

''Caveat Vendor, FTC Crackdown on Advertising for Children,'' Newsweek Magazine, 17 June 1974, p. 69.

Gaedeke, Ralph M. and others. Consumerism: Viewpoints from Business Government, and Public Interest. San Francisco: Canfield Press, 1972.

Taylor, Jack L. The Consumer in American Society: Additional Dimensions. New York: McGraw-Hill, Inc., 1974.

Time Magazine, June 14, 1971, p. 82.

Troelstrup, Arch W. The Consumer in American Society. New York: McGraw-Hill, Inc., 1974.

TYPING THE PAPER

1. Always type research papers, or have them typed for you. Use standard size type (pica); avoid very small typeface portables. Make sure your ribbon is still dark. Inexpensive computer printers can usually produce satisfactory copy if the ribbon is good or if the printer is set for "double strike."
2. Use standard typing paper only, medium weight, 20 pound bond.
   - MLA guidelines say avoid expensive, heavy-weight paper, but APA guidelines call for "heavy bond."
   - Very light computer paper should be avoided, and ruled or colored computer paper can't be used for final drafts.
   - Always remove the tractor edges (the strips with holes in them) before handing in computer paper.
   - Never use very light, see-through paper like onionskin. Never use easy-to-erase paper (it smudges).
   - Learn to use white-out liquid or correction tape for errors.
   - Type on one side of the paper only.
3. Double-space. The old style required indented quotes and footnotes to be single-spaced, but today the rule is for everything to be double-spaced. Double-space your Works Cited or References page. The only exception to the rule for double spacing is that APA style says you *may* leave four spaces (double double spaced) between the end of one section and the beginning of the next.
4. Use one of the in-text styles: MLA or APA. Ask your instructor before using footnotes or endnotes.
5. Unless told otherwise, use headings and subheadings to label the sections of your paper.
   - Center the first-level headings and capitalize the first letter of all significant words. The title of your paper, abstract, outline, notes, bibliography are first-level headings.
   - Second-level headings should be flush with your left margin, underlined, and capitalized the same as the first level. The sections of your outline identified with capital roman numerals are second-level headings.
6. Use a one-inch margin on all four edges of your paper if you are using MLA style, one and a half inch for APA. Learn to use the margin setting or line-length feature on your typewriter or computer. Standard paper is 8½ inches wide and 11 inches long. If your typewriter has no margin control, make a very light, unobtrusive pencil dot one inch from the bottom of each page and watch for that dot as you type.
7. Title pages are optional for MLA, required for APA. Ask your instructor. If no title page is desired, the MLA guidelines suggest that you put your

name, your professor's name, the course number, and the date in the upper-left (not right) corner of the first page of your paper, double-spaced. Check with your instructor if MLA is not specified. If a title page is desired, see the model research paper in the Appendix. Center the title page information. Allow six to eight lines between the title and your name and six to eight lines between your name and other information.

8. Center the title of your paper on the outline and again on the first page of the paper itself, immediately below the one-inch margin at the top of the page. All papers must have a title. Don't underline or put quotation marks around the title of your paper.
9. When submitting your paper to a professional journal, don't staple or fasten the pages in any way. For school work, most instructors want pages stapled together. It's a good idea to invest in a small stapler. Don't pin, fold, or tear corners to fasten the pages together; they will come undone anyway (and they look unprofessional). Covers of any kind are optional. Ask what your instructor's preferences are (most don't like covers). If you do use a cover (not recommended), make sure the pages are securely fastened inside so that they can't slip out. Tension binders often come apart and aren't a good choice.
10. Number your pages. In both MLA and APA styles, number *every* page, including bibliography pages.
    - Starting with the title page (if there is one), number each page with an arabic numeral in the upper right-hand corner, in both MLA and APA styles.
    - For MLA style, use your last name as a header: Smith, 1.
    - Don't write "page" or "p." or use anything other than your own name as a header in either style.
    - APA requires a running head (shortened title—not more than 50 characters) *above* each page number.
11. Use a uniform indentation for paragraphs. Five spaces is standard. APA says indent long quotes 5 spaces, but MLA says long quotes should be indented an additional five spaces (that is, 10 spaces from your left margin); it isn't necessary to indent from the right in either style. See the section on punctuation in Chapter 7. The second and all subsequent lines of bibliography entries should be indented five spaces.
12. Leave two spaces between each sentence. If you use note numbers, for example to refer your reader to a content note, no space should be left between the word or sentence in your paper and its note number. A note number at the end of your sentence comes after the period, but it's often better to word your sentence so that the note number falls within. In the "Notes" section of your paper, leave a space between the raised note number (superscript on your computer) and the first word of the note itself. (A "note" is not the same as a citation; see model "Notes" page in Appendix, p. 339.)

13. Don't use hyphens to divide words at the margins of papers you are submitting for publication (your editor will have to remove them).
    - It's a good idea not to use hyphens at all to divide words that are too long. When a word is too long to fit on the line, take the entire word to the next line—don't divide it.
    - If your computer has an automatic hyphen feature, turn it off. Typescript (typed manuscript) has no need to make the lines appear uniform in length, as they are in published material.
    - If you do use hyphens, remember to use them only between syllables; follow standard rules, and use the dictionary to check any word whose division you aren't sure about.
    - You must use *two* hyphens to indicate a dash.
14. On the whole it isn't a good idea to use "justified" lines—to make them all look the same length, like a printed page. Many printers "justify" the lines by pulling the letters and words apart, making your work more difficult to read.
15. Neither publishers nor instructors will accept papers that are hard to read. It's always wise to ask in advance about dot matrix printers, especially if you yourself are doubtful about your printer's readability.
    - APA says dot matrix printers should be acceptable, as long as each letter is dark and clear.
    - MLA says letter quality printers (whose print looks typed) are preferred.
    - You are responsible for producing professional quality work, even if that means hiring someone to do the typing for you.

ILLUSTRATIONS AND TABLES

Use visual illustrations (line drawings, charts, diagrams, and so on) or a table of numbers and statistics wherever the reader might be confused by data in your paper.

Visual material must be kept simple and clear. All visual materials should be self-explanatory; nevertheless, you must also provide a thorough explanation. Figures and charts should be carefully drawn in ink (not pencil), using a ruler for straight lines and a compass for curves. Wherever possible, type in all words and numbers: avoid printing by hand. See Figure 6.1.

Visuals should be introduced at the place in your paper where you mention them, if that is possible, or as close thereafter as practical. If there are too many such figures, or if it's impossible to position them close to your mention of them, they can become distracting. Place them at the end of your paper in an appendix if there are more than a few or if there are problems with positioning them.

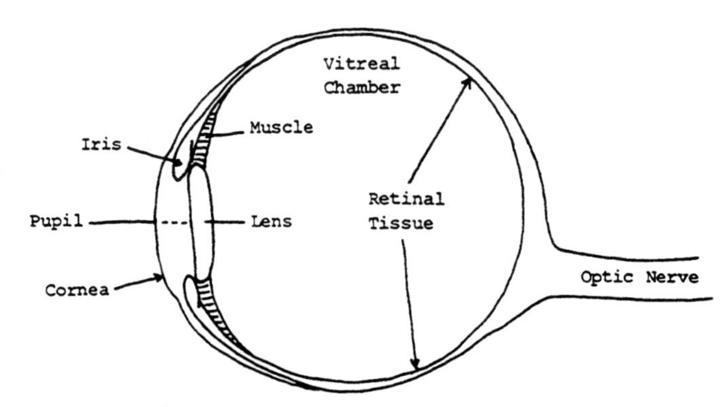

Fig. 1 Schematic Drawing of Human Eye

FIGURE 6.1 An illustration

Label and number all illustrations consecutively throughout your paper. If there are both figures and tables in your paper, use arabic numerals for the figures (Fig. 1, Fig. 2, and so on) and use full capitals and roman numerals for the tables (TABLE I, TABLE II, and so on). The label and number for a figure or line drawing of any kind are usually centered two lines below the figure, followed on the same line by its caption. The label and number for a table are usually centered two lines above the table. All illustrations and tables should have a caption in addition to an identifying label and number. See Figure 6.2.

TABLE I

Responses of 200 Business Executives to a Question Regarding the Value of Graduate Degrees to Career Advancement*

	Very Valuable	Uncertain	Of Little Value	Row Subtotal
Women	77	9	14	100
Men	58	31	11	100
Column Subtotal	135	40	25	200 Total

* Elizabeth Grigson and Ronald L. Burgess, *The Changing Marketplace: Women's Careers in Business* (Englewood Cliffs: Prentice Hall, 1982): 114.

FIGURE 6.2 A table

If the figure or table is one you found in a book or other source (not one you created yourself), it must have a footnote. The footnote should appear two lines below the table. Use an asterisk (*) instead of a footnote number, placing it one-half line above the end of the caption. If you create a chart or table using figures from a published source, use an asterisk and a footnote to show the source of the figures.

A bar graph, which is a visual presentation of data, is designated a figure, not a table. Tables are generally limited to columns of numbers.

COMMON RESEARCH ABBREVIATIONS AND TERMS

In general, avoid abbreviations and Latin terms when writing to a popular audience. However, you may use them with academic readers.

AD	*Anno Domini*, in the year of the Lord. MLA omits periods: (AD).
anon.	anonymous; the author's name is unknown. (Never appropriate for an article that is merely unsigned, as in a magazine or newspaper.)
ante	before
attrib.	attributed; authorship is not positive
b.	born
BC	before Christ. MLA omits periods: (BC).
bib.	biblical
bibliog.	bibliography
©	copyright; date of publication
ca. or c.	*circa*, about: date is approximate
cap.	capital; capitalized
cf.	*confer*, compare
ch., chs.	chapter, chapters
col., cols.	column, columns
d.	died
diss.	dissertation
ed., eds.	editor, editors; edited, edition
e.g.	*exempli gratia*, for example
esp.	especially
est.	estimated, estimation
et al.	*et alii*, and others
etc.	*et cetera*, and so forth
f., ff.	and the following page, pages

fn.	footnote, endnote
fr.	from
ibid.	*ibidem*, in the same work at a different place; cited immediately above. Ibid. means a different reference to the preceding source and should have a page number: Ibid., p. 12. Neither MLA nor APA uses ibid.
i.e.	*id est*, that is
l., ll.	line, lines
loc. cit.	*loco citato*, in the same place; same reference as before. Loc. cit. doesn't need a page number, but you must identify the reference: Smith, loc. cit. Neither MLA nor APA uses loc. cit.
MS, MSS	manuscript, manuscripts
n.	note
N.B.	*nota bene*, note well; take notice
n.d.	no date of publication
n.n.	no name of publisher (see n.p.)
no., nos.	number, numbers
n.p.	no place of publication; no publisher given
n. pag.	no pages; pages are not numbered
obs.	obsolete
op. cit.	*opere citato*, in the work cited previously, a different reference to a work cited earlier. Op. cit. requires a page number: Smith, op. cit. p. 9. Neither MLA nor APA uses op. cit.
passim	here and there; at intervals
pl., pls.	plate, plates; plural
pseud.	pseudonym, pen name
pt., pts.	part, parts
q.v.	*quod vide*, which see
rev.	revised; revision; review, reviewer
rpt.	reprint; reprinted
sec., secs.	section, sections
ser.	series
sic	thus it is; mistake in the original
v.	*vide*, see
var.	variant
viz.	*videlicit*, namely
vol., vols.	volume, volumes
vs.	*versus*, against

Abbreviations for States

Don't abbreviate state names in the text of your paper. In your Works Cited list, only the city is required unless some confusion may arise without the state. Use the official ZIP code abbreviations for state names: AL, CA, FL, IN, and so on in MLA Works Cited and APA References lists.

Abbreviations for Books of the Bible

Always abbreviate when citing chapter and verse of the bible:

I Sam. 15:14

Books of the Old Testament		*Books of the New Testament*	
Gen.	Eccls.	Matt.	I Tim.
Exod.	Song of Sol.	Mark	II Tim.
Lev.	Isa.	Luke	Titus
Num.	Jer.	John	Philem.
Deut.	Lam.	Acts	Heb.
Josh.	Ezek.	Rom.	James
Judg.	Dan.	I Cor.	I Pet.
Ruth	Hos.	II Cor.	II Pet.
I Sam.	Joel	Gal.	I John
II Sam.	Amos	Eph.	II John
I Kings	Obad.	Phil.	III John
II Kings	Jon.	Col.	Jude
I Chron.	Mic.	I Thess.	Rev.
II Chron.	Nah.	II Thess.	
Ezra	Hab.		
Neh.	Zeph.		
Esther	Hag.		
Job	Zech		
Pss.	Mal.		
Prov.			

Roman Numerals

Roman numerals for volume numbers of journals and line numbers in plays are passing out of style; you should change them to arabic in your Works Cited and References pages. Many readers today don't know the roman system, and you should use arabic numerals whenever possible. However, you must use roman numerals to cite pages that are numbered with roman numerals, such as preface pages, and you cannot change roman numerals that are part of a title. Furthermore, they were widely used to mark volume numbers on older periodicals and elsewhere, so you need to be able to read them.

Arabic	*Roman*	*Arabic*	*Roman*	*Arabic*	*Roman*
1	I	15	XV	200	CC
2	II	16	XVI	300	CCC
3	III	17	XVII	400	CD
4	IV	18	XVIII	500	D
5	V	19	XIX	600	DC
6	VI	20	XX	700	DCC
7	VII	30	XXX	800	DCCC
8	VIII	40	XL	900	CM
9	IX	50	L	1000	M
10	X	60	LX	2000	MM
11	XI	70	LXX	3000	MMM
12	XII	80	LXXX	4000	$M\overline{V}$
13	XIII	90	XC	5000	$\overline{V}$
14	XIV	100	C		

Note that when the smaller number comes first it is to be read as a subtraction from the larger number: IV = 4. When the smaller number follows the larger, it is to be read as an addition: VI = 6. A large number like 1989 shouldn't be read as 19 + 89 but as 1000 + 900 + 80 + 9: MCMLXXXIX.

STUDY GUIDE: Chapter Six Organizing Research Writing

1. What is the generally recommended order of arrangement of examples and arguments for a research paper?
2. Why not wait until data collection is finished to come up with a specific plan for your paper?
3. What is the function of an outline for a research paper? Why have one?
4. In what way is an outline "not so much a blueprint as an organic framework"?
5. What option do you have regarding the introduction and conclusion of your outline?
6. What is an abstract?
7. How long should an abstract be?
8. What guidelines should be followed concerning the title of a research paper?
9. What guidelines should be followed concerning the introduction to a research paper?
10. What are some introductory strategies for research papers? Describe a few.

11. What is the primary organizational strategy of much writing?
12. What is the advantage of specific-to-general order?
13. What is chronological order? Why use it?
14. What is spatial order? What is the difference between the two types of spatial order?
15. What is order of importance?
16. Why should a researcher make concessions in a research paper?
17. Why is it good strategy to present the weaker side first in a research paper?
18. Why should a researcher make the case for the opposition as strong as it can be made?
19. What is the chief requirement of a paragraph?
20. What is the most obvious strategy for unifying a paragraph?
21. What does paragraph development mean?
22. What is a unifying device?
23. What is the function of the conclusion to a research paper?
24. What is meant by the errors of "increasing the charge" or "extending the argument by easy analogy"?
25. What title should the bibliography page have?
26. What is the importance of a bibliography?
27. Explain selected bibliography, limited bibliography, and annotated bibliography.
28. True or false: Everything you read must be quoted or paraphrased—actually used—in your paper.
29. True or false: Anything not actually used in your paper must be eliminated from your bibliography.
30. In the bibliography, how would you list a report written by The Commission on Old Age?
31. In the bibliography, how would you list the following unsigned article: "A Projection of Statistics"?
32. In the bibliography, how would you list the following unsigned article: "45 Days Left in Semester"?
33. What things should be double-spaced in research writing?
34. Should you use a title page?
35. Should you hyphenate (divide) words that run into the margin?
36. How should figures be labeled in a research paper?
37. How should tables be labeled in a research paper?
38. How can you document a table or figure you found (versus one you created yourself)?

CHAPTER SEVEN

Style in Research Writing

Many of the variations of English that you accept as natural elsewhere are disallowed in academic writing. However, because formal writing is more orderly and predictable than natural language, there is a danger of its becoming stuffy and boring. Your task is to control both style and content; the challenge is to hold the reader's interest while writing with sources. As an example of good research writing, read the following paragraph from a scientific journal. The paragraph is about an experiment in which students pretended to be insane:

> The pseudopatient, very much as a true psychiatric patient, entered a hospital with no foreknowledge of when he would be discharged. Each was told that he would have to get out by his own devices, essentially by convincing the staff that he was sane. The psychological stresses associated with hospitalization were considerable, and all but one of the pseudopatients desired to be discharged almost immediately after being admitted. They were, therefore, motivated not only to behave sanely, but to be paragons of cooperation. That their behavior was in no way disruptive is confirmed by nursing reports, which have been obtained on most of the patients. These reports uniformly indicate that the patients were "friendly," "cooperative," and "exhibited no abnormal indications."
>
> D. L. Rosenhan, "On Being Sane in Insane Places," *Science*, 19 Jan., 1973.

The author of "On Being Sane in Insane Places" is a professor of psychology and law, writing for psychologists and lawyers. In the experiment, no doctor ever discovered that the "pseudopatients" were only faking insanity, and this fact has serious implications for both psychology and law! Rosenhan's article deals with academic subject matter and is very formal in tone and vocabulary. The writing is clear, informative, objective. Later in the report Rosenhan writes in the first person—"I do not, even now, understand the problem well enough to perceive solutions." Using the first person ("I") allows the author to avoid what many think is an awkward point of view: "*This researcher* (?) does not now understand the problem. . . ." Even with an occasional first-person reference, Rosenhan's article remains objective writing aimed at a very special audience.

However, for most writing in college, something less than extreme formality is desired. For example:

> If we were to go back into the Age of Reptiles, its drowned swamps and birdless forests would reveal to us a warmer but, on the whole, a sleepier world than that of today. Here and there, it is true, the serpent heads of bottom-feeding dinosaurs might by upreared in suspicion of their flesh-eating compatriots. Tyrannosaurs, enormous bipedal caricatures of men, would stalk mindlessly across the sites of future cities and go their slow way down into the dark geologic time.
>
> Loren Eiseley *The Immense Journey*

This is very fine writing by any criterion. Professor Eiseley is an anthropologist, but the literary quality of his nonfiction writing exerts a nearly poetic appeal. Those who believe that research writing must be nothing but the barest and dullest of factual reports have only to read more widely in the best work of modern nonfiction writers. It is through such skillful scientific writing as this that the rest of us are enticed to read outside our own areas of interest.

> This surgeon has now come to the point of threading a pursestring of stitches along the line of that clamp he clamped across the left auricle, then a second line of stitches, on the chance the first might break. (Suddenly and without looking up he spoke peremptorily to a nurse, an unanticipated severity in his voice, and one respected him only the more for that.) He sliced off the flesh above the clamp. Took away that clamp. Pushed his index finger down into that hole, into the heart. The wet tissues hugged his finger. He was inside that eighty-year-old lady's beating heart, and immediately one thought of his finger as being just there. The entire operative field danced. Deeper into the ventricle went the finger. Now he appeared almost recklessly to thrash it side to side, knew how, ripped scars, widened the passage. Then he coolly drew out the finger, replaced the clamp, washed away the debris, tightened both purse strings, reinforced with stitches all the cut layers of flesh, closed the hole in the chest. His finger had been down where the heart-sound is made.
>
> Gustav Eckstein, *The Body Has a Head*

Gustav Eckstein is a doctor, and his very fine book reveals the human body as only a doctor can know it. Certainly this is a scientific book, even a technical book, but it is aimed at the general reader, a "popular" audience. And for that reason Dr. Eckstein has taken great pains to put back into medicine that human quality modern technology so often obliterates.

You should aim for language suitable for educated audiences. That is, you should write on the formal side, but not at the extreme end of the writing continuum, where only the most specialized audiences will be able to read your work.

STYLISTIC VIRTUES

Accuracy

The chief virtue of the research style is accuracy; there is little value in research papers that are not accurate. You must try to help the reader understand, especially technical concepts, but you must be careful not to oversimplify or distort the concept in the process. For example:

> OVERSIMPLIFIED Light rays from a light microscope will not affect crystal structures.
>
> ACCURATE The photons which strike a crystal on the stage of a light microscope do not alter the crystal's position.
>
> A. Schwartz, *Chemistry: Imagination and Implication*

Clarity

Clarity is very much related to accuracy. Anything that distorts or clouds the evidence should be considered a mistake. Avoid ambiguity; unnecessary jargon; unnecessarily long, complicated sentences. No matter who your audience is, all readers will appreciate simplicity of expression. You must be careful not to oversimplify, of course, but your general rule should be always to write as clearly as you can. For example:

> UNNECESSARY AND AMBIGUOUS JARGON By two months of age the child at the developmental median will smile at the appearance of his maternal parent's face.
>
> CLARITY AND SIMPLICITY By two months of age the average child will smile at the sight of its mother's face.
>
> Atkinson and Hilgard, *Introduction to Psychology*

Conciseness

Conciseness is a great blessing to the reader. It helps the clarity of the evidence as well as limiting the reading time of your paper. Research is time-consuming enough without wordy, repetitious, redundant, loosely written sentences and paragraphs. The goal is not to write with the fewest possible words, but rather to avoid any words that can be removed to tighten, clarify, and simplify the writing.

> WORDY AND UNCLEAR Evidence for the successful termination of the experiment appeared improbable at the inception of the design stage.

> CONCISE AND READABLE From the beginning, the experiment seemed unlikely to succeed.

STYLISTIC FLAWS

Passive Mode

Excessive use of passive sentences gives writing an abstract and falsely authoritative tone. Researchers must not hide behind the passive, avoiding responsibility for their own actions, opinions, and conclusions.

> FAULTY PASSIVE The data were found insufficient. A negative conclusion was reached. The experiment was judged not to make a significant contribution to the question being studied.

> PREFERRED ACTIVE I found the data insufficient and reached a negative conclusion. It is my judgment that the experiment did not make a significant contribution to the question I had been studying.

Abstract Language

Excessive use of abstract nouns and verbs gives research writing a heavy, pedantic tone.

> EXCESSIVE ABSTRACTION *Excitation* of the central nervous system *obligates involuntary action in the area of motor activity* in the frog.

> CONCRETE LANGUAGE When an electric charge is applied to the frog's spinal cord, the frog jumps involuntarily.

Vague Language

Vague, imprecise, or falsely metaphoric language is usually the result of careless thinking: unclear, inaccurate, and wordy all at once.

> VAGUE The rat very often seems almost to anticipate the stimulus with extreme desire.

> PRECISE After the first series of stimuli, the rat no longer has to be forced to touch the lever but goes immediately to it and presses it repeatedly until the

> current is turned off. Rats that are allowed to continue the stimuli indefinitely push the lever until they die of exhaustion.

Sexist Language

Sexism is inappropriate in research writing. Avoid a masculine bias in pronouns or other words that indicate gender. Write in the plural: "*Researchers* will find *they* must use the procedure with care." Avoid inappropriate use of *he, him, his*: "The researcher will find that *he* (?) must use the procedure. . . ." Substitute gender-neutral equivalents for words with male connotations: *humankind* instead of *mankind*, *spokesperson* instead of *spokesman*, and so on.

Inappropriate Verb Tense

Write in the past tense:

> I placed [not *place*] the mice in the box and turned [not *turn*] on the light.

The past tense is generally correct to preserve the sense that you are reporting research already completed. However, there are some exceptions to the general rule. You may sometimes address the reader in the present tense, as if explaining your reasoning or speculating aloud:

> If we [meaning the reader and you] *see* that the light rays *are* distorted, what *does* this mean? What *causes* this distortion?

Furthermore, it is customary to identify source material in the past tense but to treat its substance in the present tense:

> Shakespeare *wrote Hamlet*. *Hamlet* is [not *was*] a fine tragedy. In the play, Hamlet kills [not *killed*] Polonius.
>
> The Rogers study *is* old. It *was written* in 1921. Rogers suggests [not *suggested*] a new direction for genetic research.

Because the material still exists in time with us, we speak of it in the present tense, even though it was written long ago. When some events occur further back in time than others, use the past perfect tense for the earlier event:

> We opened the cylinders and discovered that the seals *had leaked* and the gas *had evaporated*.

ACTIVITY 44 Edit the following excerpt for style. Change or remove anything you think is not good style. Feel free to add anything that will improve the style of this excerpt from a student paper.

Special Bulletin

The majority of Americans have become informed about world-wide events going on in the world through assimilation of television news. On the evening of March twentieth, however, the information which was received by the American public via the NBC TV movie Special Bulletin wasn't the real news at all. Special Bulletin is shot entirely on video tape, simulational of the format of a fictitious television network's (R.B.S) news coverage of a fictional crisis. The gimmick is reminiscent of Orson Welles' adaptation of H. G. Wells' War of the Worlds on radio. NBC makes much ado by way of providing plenty of disclaimers with which to make sure the viewer was aware that he was viewing a fantasy, but this doesn't detract from the realistic quality achieved by Special Bulletin while poignantly addressing the issue of nuclear arms build up.

The premise of the plot itself, which duplicates the way in which real life networks would cover a major crisis, is a novel concept that isn't like anything you'll see on commercial television this season—except of course any real special bulletins. George Bullard, who is the television critic for the Detroit News gives a summation of the plot and its suspenseful conflict as ''the approach of a deadline for the government to act on terrorists' demands that it destroy a stockpile of nuclear bomb detonators. If the U.S. doesn't comply, the terrorists threaten to detonate a nuclear device that they have built especially for the occasion. They kidnap a TV reporter and cameraman to relay their demands via a special national TV hookup, provided by an unreal network called R.B.S.'' (Bullard). The network emulates the story as a special news bulletin after the kidnapping, and perseveres with the story until its climactic conclusion is reached.

The reporting of the event by R.B.S. successfully captures the reality of real life network news by adopting the same quick pace employed by networks to obtain every angle and facet of the story. R.B.S. constantly jumps from place to place using the same techniques which are seen on the three real networks. From slick graphics, to mini-cams, and back to updates in the newsroom in Washington, and on the tugboat with the terrorists, they cut with dexterity. Even feature stories such as background exposés on the terrorists and interviews with residents of the endangered town are included.

As R.B.S. switches to its on-the-scene reporter, Meg Marclay (played by Roxanne Hart) after the nuclear explosion, the director of NBC uses his very good special effects to underscore

```
the tragedy. The dark and gloomy atmospheric lighting resembles
the overwhelming ash that pervades after a mushroom cloud filters
through the air like a death note on a trumpet. A video tape that
recounts the shocking brightness of the event is then replayed
by Meg, herself overwhelmed by the explosion.
```

PUNCTUATION IN RESEARCH WRITING

Punctuation is an important aspect of style in research writing. It is the means by which you tell the reader how to interpret your sentences. Accuracy of expression is essential, and it is therefore important to punctuate carefully. The rules for punctuation in formal research tend to be more traditional and conservative than in other kinds of writing: There is less room for experimentation with style. Options that might be acceptable in less formal writing are generally not used in research writing, especially when they increase the chances for misreading.

PERIOD

Use a period at the end of a complete sentence. Do not use a period or a capital letter for a sentence inserted into another sentence:

```
The Nigerian team protested frequently—their methods were
never questioned—that theirs had been a most rigorously con-
trolled project.
```

When an abbreviation or initial comes at the end of a sentence, no additional period is required.

```
Her real name is Harcourt Clarissa Lederer, but she is known
affectionately as H.C.
```

The initials of well-known organizations may be used without periods, after the name has been given in full once: Central Intelligence Agency (CIA).

QUESTION MARK

Use a question mark for direct questions. A question inserted into a statement retains its mark.

```
Thus the question which you put to me—what is to be done to rid
mankind of the war menace?—took me by surprise.

                    —Sigmund Freud, Letter to Albert Einstein
```

Use a question mark to question each item in a series.

```
Is it possible for the air pressure to remain steady at 100 feet?
1000 feet? 5000 feet?
```

EXCLAMATION POINT

The exclamation mark indicates surprise or strong emotion. The mark is rarely used in research writing.

COMMA

Use a comma to separate two sentences joined by *and, but, or, nor, so, for, yet*:

```
The experiment was finished, but we must still analyze the data.
```

Use a comma after introductory words and transitional expressions:

```
After her lecture, the professor answered questions from stu-
dents.
```

Use commas to separate items in a series:

```
The lab apes had very few requirements: food, water, and a stout
cage to hold them.
```

In research writing, *accuracy* is the foremost consideration. Therefore, most researchers feel it is a mistake to omit the comma after the next-to-last item in a series.

```
The refugee population was composed of people whose ancestry was
Vietnamese, Laotian, Thai, Cambodian and Chinese.
```

What does the researcher mean here? How many groups are identified? To avoid any misinterpretation, follow the custom of using a comma with each item in a series. Avoid writing sentences that can lead to misunderstanding. If the researcher really means to identify four groups, he or she should write the sentence to make that clear:

```
The refugee population was composed of Vietnamese, Laotians,
Thai, and Chinese Cambodians.
```

Use commas to set off explanatory and parenthetical elements:

```
The explosion, to put it mildly, brought an end to our lab work.
```

Use commas to set off appositives. Appositives are words that identify or rename a preceding noun or pronoun.

```
And this man, John Glenn, had given them an answer as sentimental
as the question itself.

                                 Tom Wolfe, The Right Stuff
```

Use commas to set off nonrestrictive modifiers. Nonrestrictive modifiers offer extra information but do not serve to limit or specify the word they modify. Such nonessential modifiers should be set off with commas.

```
His text, which covered more than twelve double-spaced pages,
concluded with a quotation from the Bingham censure case.

                              Theodore C. Sorensen, Kennedy
```

Do not set off restrictive modifiers. The restrictive modifier is necessary to tell the reader how many or which one is being discussed.

```
Here was the cerebrum of a European intellectual who had achieved
momentary renown before fading into the obscurity of this dusty
shelf.

                                 Carl Sagan, Broca's Brain
```

Use a comma in place of *and* between movable adjectives. If all the adjectives describe the same word, so that you could rearrange them in some other order, separate them with commas.

```
The key instrument was a long, hard, cold knife.
```

SEMICOLON

Use a semicolon to connect two closely related sentences:

```
Pecuniary power expresses itself in highly unstable form; it
offers financial reward for conformity or threatens financial
damage for dissent.

             John Kenneth Galbraith, The New Industrial State
```

Use a semicolon in a series between items containing commas.

> Scientists arrived from such places as Nome, Alaska; Chicago, Illinois; Paris, France; and Venice, Italy.

Use a semicolon in a compound sentence. Compound sentences joined by coordinate conjunctions usually need a comma before the conjunction. However, when the sentences are long or there are other commas in the sentences, you may substitute a semicolon for the comma before the conjunction.

> The spoil-sport is not the same as the false player, the cheat; for the latter pretends to be playing the game and, on the face of it, still acknowledges the magic circle.
>
> John Huizinga, Homo Ludens

Use semicolons for multiple references in one note; (Ezekiel 11: 9–13; 12: 4–5; 16: 7, 9, 25)

COLON

Use a colon to introduce a series.

> Missing from their diet were the following: calcium, sodium, niacin, and certain amino acids.

The colon introduces a series when there is a pause and a clear signal such as "the following," or "as follows." Without one of the signals, the colon should be used only when it takes the place of the expressions "such as" or "namely."

> Implied ''such as'' or ''namely'':
> Naturally, they're carrying every piece of garbage imaginable: the folding aluminum chairs, the newspapers, the lending-library book with the clear plastic wrapper on it, the sunglasses, the sun ointment, about a vat of goo—
>
> Tom Wolfe, The Pump House Gang

When "such as," "namely," or "like" is expressed (versus implied) no colon should appear:

> They were fond of drinks like margaritas, mai tais, and wallbangers.

Use no colon when there is no break or signal for a series:

> When we loaded, we had to tear the paper with our teeth, empty a little powder into a pan, lock it, empty the rest of the powder into the barrel, press paper and ball into the muzzle and ram them home.
>
> Otto Eisenschimal and Ralph Newman, Eyewitness: The Civil War As We Lived It

It is not formal style to use a colon after forms of the verb *to be* (am, is, are, was, were, be, being, been):

Informal style (not recommended for research)

> Missing from their diet were: calcium, niacin, and certain amino acids.

Formal style (preferred)

> Missing from their diet were calcium, sodium, niacin, and certain amino acids.

Use a colon before an example, illustration, explanation:

> Its effect on the process was dramatic: as the pressure increased, people began to faint.

A quote given as an illustration or example should be introduced with a colon. (All quotes require documentation.)

> It is understood that Frennel's attitude is biased: ''No other laboratory in the world can match Frennel equipment, Frennel staff, or Frennel results'' (Headley 97).

PARENTHESES

Use parentheses to set off clarifying information and information not grammatically connected to the sentence:

> The tables for oxidation rates (see p. 15) indicate an increasing rate over extended periods of time.
>
> Robert S. Richardson, The Fascinating World of Astronomy

It is possible to have one parenthetical element within another; use dashes for the first break and parentheses for the second:

> These conditions-graphed in government publications (see Appendix B)-exist mainly in the Northwest.

Use parentheses to list or outline within the text:

> This view does not take into account (1) the possibility of redundant function; and (2) the fact that some human behavior is subtle.
>
> Carl Sagan, The Dragons of Eden

DASH

In typing, make a dash with two hyphens (--) and leave no space before, between, or after them. Never use a single hyphen for a dash:

> She set the chemicals out on the counter-tables would come later when the lab was fully equipped. [The single hyphen makes ''counter-tables'' look like a hyphenated word, and produces misreading.]

Questions or exclamations inserted between dashes should retain their question marks or exclamation points. (The period would never be appropriate.)

> We all sat there—on the Hellespont!—waiting for it to get light.
>
> Ronald Blythe, Akenfield: Portrait of an English Village

HYPHEN

Typed manuscript should avoid the use of the hyphen to divide a word at the end of a printed line.

Use hyphens in compound words.

> Vice-president   brother-in-law   one-half

When in doubt, check your dictionary. As a general rule, hyphenate any compound word used as an adjective before a noun: *well-written paper*. This rule does not apply to *-ly* words, which are never hyphenated: *roughly cut diamonds*, but *rough-cut diamonds*. Modifiers that are hyphenated when they precede the noun do not need hyphens when they come after the noun: *Your paper is well written.*

Use a hyphen in words with certain prefixes and suffixes: *all-knowing, half-developed, self-serving, president-elect.* When *self* is the root word, it is not hyphenated: *selfhood, selfish.*

BRACKETS

If your typewriter or computer does not have brackets, you should draw them in with a pen. Use brackets to set off clarifying material you insert into quotations.

> By autumn the poet [Poe] was again destitute and Mrs. Clemm now exerted herself to secure him some salaried work.
>
> Hervey Allen, The Works of Edgar Allan Poe

If material you quote contains an error made by the original author, you may insert "sic" in brackets immediately after the error. "Sic," which means "thus," indicates to your reader that the error in logic or language appears "thus" in the original.

MLA Style

> Heston reports that ''10,00 [sic] people were given the serum over a period of days.''

Note that "sic" is not an abbreviation and does not require a period. It is not underlined (not set in italic print) in the *MLA Handbook*, 3rd edition, 1988; but if you are using APA (name and date) documentation, you must underline *sic.*

If your use of "sic" falls outside the quotation, it should be set off with parentheses instead of brackets: The head of the APA was once introduced as "a famous psychotic" (*sic*).

UNDERLINING (ITALICS)

Manuscript you submit for publication should use underlining for italics even if your machine has italic type. For papers in school, italic print instead of underlining may be acceptable; check with your instructor.

Underline the titles of books, booklets, pamphlets, magazines, newspapers, long poems, plays, albums (records, tapes, or cassettes), operas, films, filmstrips, works of art, and the names of radio and television series. Do not underline titles of documents like the Constitution or the Declaration of Independence (which are not considered publications).

> Beatniks and other eccentrics are discussed in Lawrence Lipton's book The Holy Barbarians.

Court cases are underlined in the text of your paper but not in your bibliography. If you are using MLA documentation, do not underline the abbreviated form of *versus* in your paper: *Sullivan* v. *The New York Times*. In APA style, court cases in your text look like this: *Carmody v. Paine* (1987) or (*Carmody v. Paine*, 1987).

Use underlining for the names of ships, planes, and trains.

> The Enterprise is the largest man-made vessel in space.
>
> Whitfield and Roddenberry, The Making of Star Trek

Use underlining for foreign words and phrases.

> What increased the danger was that at first we mistook their wild howls for cries of Vive l'Empereur!
>
> Count Philippe-Paul de Segur, Napoleon's Russian Campaign

APOSTROPHE

Add *'s* to form the possessive of any singular noun, including those that end in *s*: *Watson's hat, Mr. Jones's house, corpse's hand, iris's color.*

Add *'s* to the *plural form* of the word, unless the plural ends in *s*, in which case just add the apostrophe: *policemen's whistles, children's shoes, dogs' tails, the Joneses' house.*

QUOTATION MARKS

Use quotation marks for titles of short works. The titles of most poems, short stories, magazine and newspaper articles, book chapters, specific episodes of continuing television and radio programs, popular songs, and specific works on albums (record, tape, or cassette) require quotation marks: "The Raven" is a well-known poem by Poe.

Use quotation marks around words referred to as words.

> Your pronunciation of ''nudnik,'' by the way, is appalling. It's ''nudnik,'' not ''noodnik.''
>
> S. J. Perelman, Writers at Work, Ed. George Plimpton

Where the subject of your research is language, or when there are many references to words in your paper, use underlining (italics) instead of quotes.

Use quotation marks around all direct quotes.

> Arthur replied, ''God is total awareness.''

Use quotation marks around source material incorporated into your own sentences. Incorporating a quote does not create the need for a comma.

```
We knew she was one of our finest young artists.
We knew she was ''one of our finest young artists.''
```

If the sentence could not require a comma or colon without the quotation marks, it will not require a comma with them. Do not use quotation marks for long quotes. In MLA style, you should indent (set off or block or display) any quotation that takes up more than four lines of your text. In APA style, set off any quotation longer than 40 words. Indent all lines of the quotation ten spaces from the left margin (APA says 5 spaces) and, like the rest of your paper, type the quote double-spaced. The right-hand margin need not be indented. Copy the quotation exactly as you find it, including any quotation marks in the original. The parenthetic reference comes after the last sentence period of both MLA and APA long quotes.

Indented Quote

```
Our democratic society is clearly divided into elite and sup-
porting classes, according to Packard:
          Our class system is starting to bear a resemblance to
          that which prevails in the military services. In the
          service there are, of course, status differences be-
          tween a private and a corporal and between a lieutenant
          and a captain. The greatest division, however, is be-
          tween officers and enlisted men, with only quite lim-
          ited opportunities for acquiring, while in service,
          the training necessary to pass from one division to
          the other. (Packard 38)

                        Vance Packard, The Status Seekers
```

Use quotation marks with other punctuation. Commas and periods always go inside the quotation marks. This rule applies regardless of how many quote marks are involved. Colons and semicolons always go outside the quotation marks. Question marks and exclamation points go inside or outside the quotation marks. If the matter inside the quotes is a question or an exclamation, the mark goes inside. If the matter inside the quotes is not a question or an exclamation, but the rest of the sentence is, the mark goes outside.

```
The letter asked, ''Do you need delivery before April?''
What do you mean, ''I never sent the invitations''?
```

Use single quotes to indicate a quote within a quote.

> In her notes I found this reference, ''I've spent half the night reading Peller's article, 'Dormant Inhibitors.'''

ELLIPSIS POINTS

Use three ellipsis points to show that you have omitted one or more words from quoted material.

Quoted Matter

> ''Tampering with another nation's satellite would be regarded as a hostile act with serious consequences, and would only be done in a wartime emergency.''
>
> Frank Trippett, ''Milk Run to the Heavens.'' Time 12 Jan. 1981: 10–14.

Quoted Matter with Ellipsis

> ''Tampering with another nation's satellite . . . would only be done in a wartime emergency.''

Leave a space before and after each of the three points. If the ellipsis ends the sentence, the sentence period becomes a fourth point. The four ellipsis points at the end of a quote may represent part of a sentence or several sentences omitted.

> ''Tampering with another nation's satellite would be regarded as a hostile act with serious consequences. . . .''

SLASH

Use a slash (solidus) to mark lines of poetry in running text. One or two lines of poetry may be run in (incorporated into your text instead of set off). Use a slash to separate run-in lines of poetry.

> Crane says, ''I stood upon a high place, / And saw, below, many devils / Running, leaping. . . .''
>
> Stephen Crane, ''I Stood Upon a High Place''

Use a slash to mark inclusive dates and fractions.

> (Jan./Feb 1989)   1/2, 3/4

Use a slash to mark grammatical choices.

```
and/or   either/or
```

CAPITALIZATION

Use formal capitalization throughout research papers. Use formal capitalization for titles of publications, even if the title page uses unusual capitalization (such as all uppercase letters). In general, capitalize the first word and the last word and all significant words in between (excluding articles—*a, an, the*—coordinate conjunctions—*and, but, or, nor, so, for, yet*—and short prepositions—*at, by, below, from, in, on, to, up, with*).

Capitalize the first word of a subtitle following a colon.

```
Hydrotherapy: A Study of Arthritis
''Camping: The First Day''
```

Do not capitalize *the* as part of a newspaper title in your paper:

```
the New York Times     the Christian Science Monitor
```

Note that *The* is usually dropped from footnote and bibliographic references to newspapers and journals, even if it appears as part of the title on the periodical itself.

ACTIVITY 45 Edit the following sentences for punctuation. Remove extraneous marks, add missing marks. There can be more than one way to punctuate; feel free to edit any way that improves the sentence.

1. The lab technician said I do not think there is any cat food here.
2. A child shouted They are giving away free ice-cream at Delaney's!
3. Please put away your things and come with me said the principal.
4. We have no need to buy more spaghetti replied Mrs. Entwhistle.
5. I think said Ollie these are the girls who ordered our new giraffe collars.
6. You think you are so smart said Nadine. Well, you're not. I think you're quite limited intellectually.
7. The book said communism is: "based on the systematic murder," of the middle and upper classes.
8. We had been studying man's 'primordial' evil.
9. Lucia said, "I don't date guys who say, "stinky feet."
10. Professor Longbarrel said, We will read Poe's "Pit and the Pendulum" next week.

11. Corky asked, "Professor Twaddle, didn't you say, "Don't skip anything in the book?"
12. We watch "Dallas" all the time, and we especially enjoyed the episode called, *Who Shot J.R.?*
13. Invictus was one of my favorite poems in the eighth grade.
14. This morning's *Free Press* has an article called Crooks Catch Cow!
15. The young professor began her class by announcing, "I have a Ph.D.
16. Hogan shouted across the library, "How do you spell Shakespeare?
17. Miss Oglesquint asked, "Who has read the short story, 'The Open Window'"?
18. Our equipment was delayed in Norway, nevertheless we decided to go on with the experiment.
19. My reader says I have too many ands in my paper.
20. The corpse long stuck on the bottom of the brine tank was finally pulled up with a common boat hook.

STUDY GUIDE:
Chapter Seven Style in Research Writing

1. What is style?
2. What is the challenge of style in research writing?
3. Should researchers use the pronoun "I"?
4. What is "popular" writing?
5. What is the chief virtue of research writing?
6. What is jargon?
7. What is concise writing?
8. What is a passive sentence? What is wrong with using the passive?
9. What is abstract language?
10. What is sexist language?
11. What is the appropriate verb tense for most research?
12. What is the significance of punctuation?
13. What is an appositive?
14. What is a nonrestrictive modifier?
15. How is a dash indicated on a typewriter or computer?
16. What are brackets used for?
17. What kinds of titles should be underlined?
18. What kinds of titles should be placed in quotation marks?
19. What is an incorporated quote?
20. What is an ellipsis?
21. What is the rule on capitalization in titles of publications?

APPENDIX

Writing Situations

Research writing in the sciences and in the humanities can be formal or informal; it can be concrete and experimental or philosophical and theoretical. What governs this diversity is the writer's situation. When writers perceive a purpose and an audience for a given subject, they write whatever seems appropriate to the situation. In general, the subject matter tends to be scientific or academic, the audience is usually perceived as intelligent and educated, the writers tend to see themselves as information givers, and their general purpose is to describe reality. Nearly all research writing is based on the principle of substantiation; evidence is presented in substantiation of a thesis. (See p. 8.) Once you learn this basic principle, you can apply it to many types of research assignments.

WRITING SUMMARIES AND ABSTRACTS

Summarizing is a useful skill for researchers. If, for example, you wanted to collect information about the AIDS virus, one way to do it would be to summarize articles in the library. In this way you can build up a supply of condensed information useful for your project. If they are done well, you can often learn as much from a good summary as from the full-length original.

Finally, summarizing is an excellent study device. In order to assimilate your reading it's necessary to absorb what you have read. Summarizing permits you to interact with your reading, make judgments about what is important (and must remain) and what is less important (and can be dropped). The ability to visualize information in condensed form is a key skill for researchers.

Summary

A summary is a condensed version of a text. Depending on your purposes in summarizing, the summary can be loose and informal or very precise and accurate. Summaries can be any length that suits your purpose. It is possible to give a very "summary" overview of an entire book in a single sentence:

> Moby Dick is the story of Captain Ahab's pursuit of the great white whale, Moby Dick, that once bit off the captain's leg.

The point is to shorten, condense, abbreviate the original. Various researchers suggest lengths of one third to one fourth of the original length. The one-page summary is very popular. The 250-word abstract is nearly a standard. In lieu of any other directions (ask your instructor), the one-page or 250-word limit is recommended, whichever is shorter.

Descriptive Summary

A descriptive summary is useful for calling attention not only to the information in an article or book but also to the author's techniques. It is also useful in those situations in which it is important to keep clear which are your own words and which are those found in the original. For your own use this kind of summary can be very helpful in recalling significant features of the author's form and style.

In the descriptive summary, the point is to describe the original article. It is appropriate to mention the author's name throughout, to comment on the author's strategies ("Hensen refutes his critics with data from his own studies. This report discusses a study. . . . The study examines three possibilities. . . .") and style and to quote the author's words where appropriate:

> The author resorts to a question-and-answer style: "Has anyone actually seen a retrino? Bailey reports one in his 1985 study."

Objective Summary

An objective summary requires that you add nothing of your own. Such a summary sounds almost like the original. It's as if the original author had decided to rewrite his/her article to shorten it greatly. In an objective summary, make no references to yourself, to the title, to the article itself, the author, or the author's strategies and techniques. You should write your summary as if you were the original author, rewriting your article. It shouldn't be possible to tell that what you have written is a summary. That is, anyone not familiar with the original shouldn't receive any clue from you that you are summarizing.

Abstract

The abstract is a standard component of scientific writing. Doctoral dissertations require a 600-word abstract for *Dissertations Abstracts International.* Scientific journals usually specify a 250 word abstract. Many APA journals require an abstract of 100–150 words. Abstracts serve everybody's interests: They make the researcher examine his or her work in the cold light of objective analysis, stripped of everything but its bare essence. An impressive 10- or 20-page manuscript may prove to be less impressive when reduced to its essential message. Journal and book editors need a quick, accurate look at your work before plowing through a lengthy manuscript. Finally, your readers will benefit. For the casual reader, your abstract serves as a mini-preview. For the serious researcher, your abstract will show whether your work is relevant.

Your abstract should be as detailed as you can make it in the space you have. Ideally, the abstract will have everything in it that your paper has; but practically speaking, you must get at least all the key points of your research into your abstract. The abstract of an experiment should contain at least the hypothesis being tested, the methodology, and the results.

The Process of Writing a Summary

Read Carefully

Read first for an overview of the work; make sure you understand the writer's purpose. Look up unfamiliar words, references, and figures of speech.

Read a Second Time, Taking Notes

On the second reading, find the author's thesis and mark it or write it in your notebook. Once you have identified the thesis, find the major divisions of the work; look for topic sentences and controlling ideas in paragraphs and sections.

Make an Outline

Analyze the structure of the work. Make yourself an informal outline.

Write a Preliminary Draft

1. If you are writing an objective summary, write as if you were the author. Make no references to yourself (as author); add no information to the summary that wasn't in the original. If you are writing a descriptive summary, write in the third person: "The author says. . . ."
2. Give as much information as you have room for. Make sure you get the main points, but also include as many of the details and examples as you can. The ideal summary contains, in condensed form, all the information found in the original source.

3. The 250-word abstract is standard in science—about one page of double-spaced typing. However, some works are very concise and not easy to reduce without cutting something important; others are more loosely written and can be greatly reduced. Therefore it isn't easy to generalize about the length of a summary. Be guided by your audience—who you are summarizing for—and your purpose.
4. Avoid distortions of the original. Your summary should reflect the author's point of view as well as the original structure and development. If the author is mostly concerned with the storage of nuclear waste and only minimally interested in nuclear terrorism, don't treat the two ideas as equal in your summary.
5. In any summary, whether descriptive or objective, use your own language as much as possible. Don't write your summary by lifting topic sentences out of the author's paragraphs. You shouldn't end up with as many paragraphs in your summary as the author has in the original. Use your own words to rephrase and condense the ideas in the original. Wherever you do use the author's words, put them in quotes.

Revise the Draft

Check your draft against your outline and against the original. Does it cover the points in the original? Do you have room for additional information? (If you are permitted to have an entire page, 250 words, use all the space you have.) Does your draft sound like the original—does it have the same tone? Is your draft readable? The requirements of summarizing shouldn't be used as an excuse for poor writing. Your summary should be a readable, fluent, coherent whole, not a choppy collection of disconnected ideas. Anyone can make a list of points, but it takes skill to produce a readable summary. Use transitions. Avoid repetitions; condense your sentences:

> REPETITIOUS
> The author favors national health insurance. The author also favors hospital vouchers for the poor. Then, too, the author believes a program of preventive medicine should start in the schools.

> CONDENSED
> The author favors national health insurance, hospital vouchers for the poor, and a program of preventive medicine starting in the schools.

Check your draft and remove any of your own personal intrusions, opinions, or editorial comments. The promise of summary writing is that we will get the same information in fewer words, that the summary means the same as the original, but is shorter. You must not permit your approval or disapproval of the original to show in your summary.

Read the following article. What are its main points? How would you summarize it?

AN EXPLANATION OF THE LANGUAGE OF A CHIMPANZEE

We have been studying the LANA Project of Rumbaugh and his colleagues (1) with two main objectives: (i) to describe the minimum set of abilities necessary to account for the chimpanzee's use of the Yerkish keyboard language and (ii) to examine the extent to which principles of reinforcement can explain this behavior. We propose that Lana's behavior can, to a large extent, be attributed to two basic processes: paired associate learning and conditional discrimination learning. Lana was capable of rapidly learning a large number of paired associates (that is, lexigrams appropriate for particular objects, people, or events). She was also able to produce one of several ''stock sentences'' of lexigrams depending upon situational cues, a conditional discrimination. In producing these sentences she was able to insert an appropriate paired associate in an otherwise fixed string of lexigrams.

Table 1 is a decision table of conditions that we propose determine which of six stock sentences is appropriate in a given context. The table shows the circumstances under which each of the sentences was used. These are whether an experimenter has an object, whether it is food, and if so, whether it is placed in a dispensing machine. For example, if food is in the machine, the first stock sentence (S1) is used [for example, (Please machine give milk.)]. In contrast, if a nonfood object is present, then S2 is used [for example, (Please Tim give ball.)] Table 2 lists paired associates for the people and incentives that are part of a given text.

All of the sentences in Table 1 share the following features: constant elements (for example, the lexigrams for starting and ending a sentence), variable elements (for example, the name and incentive), and finally, an element for the activity (such as ''give'' or ''move''). A

TABLE I

On the basis of yes (Y) or no (N) answers to the three questions, a stock sentence is chosen. For example, if food is present but not in the machine S2 (stock sentence 2), S3 or S4 is chosen probabilistically. In these cases one of two activity elements is chosen probabilistically, and a paired associate for the person and incentive present is substituted for the variable ‹name› and ‹incentive›. See table 2 for the lists of paired associates.

Queries	Answers			
Object present?	Y	Y	Y	N
Object food?	Y	Y	N	
Object in machine?	Y	N		
Stock sentences				
S1: Please machine give ‹incentive›	+			
S2: Please ‹name› {give / move into room} ‹incentive›		+	+	
S3: Please ‹name› {move behind room / put in machine} ‹incentive›		+		
S4: Please Lana {want eat / want drink} ‹incentive›		+		
S5: Please ‹name› {tickle / groom / swing / carry out-of-room} Lana				+
S6: Please machine make ‹event›				+

prototypical stock sentence can be represented as (Please ⟨name⟩⟨activity⟩⟨incentive⟩.) Lexigrams in the same semantic class have the same color, but the sequence of colors is not the same for all sentences. In some cases a stock sentence involves additional lexigrams (''out-of-room'' and ''in machine'') whose selection is determined by the particular activity word chosen. For example, in S2 selection of the activity word ''move'' entails selection of additional words (''into room'') to be placed at the end of the string. In Table 1 an activity word and its additional words appear together to show this relationship.

In order to test this analysis of Lana's behavior, we wrote a program in BASIC, building into the program the set of skills described above. Input to the program consists of a description of the context as shown in the sample output of the program in Table 3. The data base consists of the six stock sentences listed in Table 1 and the paired associates listed in Table 2.

The program works as follows. The answer to the first three input questions determine which of the six stock sentences is employed; the answers to the last three input questions determine what value the variables of the sentence will take. For example, if a food object is present in the machine, S1 will be chosen (Please machine give ⟨incentive⟩.) If the object is milk, the appropriate lexigram for milk would be substituted for the variable

TABLE II

Semantic Classes of Paired Associates.

	Incentives					
Names	Foods*	Unit Foods	Liquids	Objects	Unnamed	Events
Tim	Apple	M & M	Coffee	Ball	This	Movie
Beverly	Banana	Nut	Coke	Blanket	which-is	Music
Bill	Bread	Raisin	Milk	Bowl	<color>†	Slide
Visitor	Cabbage		Juice	Box		Television
You	Chow		Water	Can		Door-open
Lana				Cup		Window-open
Me				Feces		
				Shoe		

* In S1, words from this class are always preceded by the lexigram ''piece-of.''

† Colors: black, blue, green, orange, purple, red, white, and yellow.

⟨incentive⟩, and the sentence would be (Please machine give milk.). In some cases, more than one stock sentence is possible in a given context. For example, either stock sentence 5 or 6 is appropriate if no object is present. In these cases, the program selects one or the other in a random manner. In some cases, more than one activity word is appropriate to a particular sentence type. This choice is also made probabilistically by the program. For example, the program selects for S5 one of the four possible activity words tickle, groom, swing, or carry out-of-room.

TABLE III
Sample Results from BASIC Program Developed to Simulate Lana's Behavior.

Context		Paired Associates		
Query	Answer	Referent	Lexigram	Sentence
Is an object present?	Yes	Who is the person?	Tim	Please machine give milk.*
Is the object food?	Yes	What is the object?	Milk	
Is it in the machine?	Yes			
Is an object present?	Yes	Who is the person?	Beverly	Please Beverly move ball into room.†
Is the object food?	No	What is the object?	Ball	
Is an object present?	Yes	Who is the person?	Tim	Please Tim move this which-is yellow into room.‡
Is the object food?	No	What is the object?	Pencil	
		What is its color?	Yellow	
Is an object present?	No	Who is the person?	Jane	Please visitor groom Lana.§

* When the answer to all three conditional questions is ''yes,'' S1 is selected, with substitution of the paired associate for the incentive present. † When an object is present, but is not food, S2 is selected. In this case, the activity word selected was (move + into room). ‡ The context is similar to that above, but no paired associate has been stored for the object present. § When no object is present, S5 or S6 is chosen probabilistically; this is S5. In this example, no paired associate has been stored for the person present.

The model that has been presented is based upon the 198 sequences that appeared in descriptions of Lana's behavior in Rumbaugh's book (1). After the computer program was developed, we were given the data from the appendix of a thesis by Gill (2), which contained descriptions of all contents in situations involving conversations with Lana as well as complete protocols of Lana's productions during a 3-month period centered about the time of the conversation experiments. We have analyzed all of the data in Lana's protocols for the period of the conversation studies, and selected at random the protocol for 1 day per week for each of 5 weeks preceding and following the thesis period.

The results of the analysis are summarized in Table 4. We have divided Lana's productions into three main classes: (i) stock sentences (one of the six sequences in Table 1), (ii) nonstock sentences (other sequences that were reinforced), and (iii) errors (sequences that were not reinforced).

Table IV
Analysis of Lana's Productions.

	Frequency of Occurrence with Experimenter	
Type of Production	Present	Absent
Stock sentences	5725	5288
Trained nonstock sentences	766	73
Novel nonstock sentences	500*	1
Errors	1663	467

* Excluding errors (nonreinforced responses) the model based on 198 sequences in Rumbaugh's book (1) fails to account for 6 to 8 percent of the larger sample of 6991 sentences (1 percent confidence interval).

When no experimenter was present, 91 percent of all productions (5288 of 5830) were one of the six stock sentences, 8 percent were errors, and only 1 percent were nonstock sentences. The model accounts for the stock sentences; it does not attempt to account for Lana's errors. The nonstock sentences, however, are exceptions to the model. Of the 74 nonstock sentences that occurred, 73 were the sequence (Please machine give piano.), which is identical to our S6 except that the activity word for obtaining piano is give instead of make. This represents no challenge to the model since it could easily be expanded to include one additional stock sentence.

For sequences produced when an experimenter was present, our model is less successful. Of all productions 66 percent were one of the six stock sentences, 19 percent were errors, and 14 percent were nonstock sentences. Again, the nonstock sentences represent exceptions to the model, and therefore we considered them in some detail. Of the 1276 such sequences that Lana produced when an experimenter was present, 776 of these were specifically trained sequences such as (⟨object⟩ name-of-this.). This leaves 500 novel sequences to be accounted for. Twenty-eight of them were fragments such as ''box orange'' that would normally have been nonreinforced. All of the remaining sequences included the activity words give, put, move, want, or make. While these sequences constitute exceptions to the six stock sentences as we have presented them, they showed many of the same elements and sequencing relations as the prototypical stock sentences shown earlier and seem to represent extensions of it, rather than departures from it. Lana produces such sequences when she has not been able to obtain an incentive in the usual way with a stock sentence.

The evaluation of the analysis required complete and accurate records of a large corpus of productions. Such data were collected by Rumbaugh (1977) since the lexigram board was interfaced to an on-line computer. Terrace *et al.*(3) also objectively analyzed a large corpus (35 hours of videotape) of productions and their contexts of the chimpanzee Nim. The abilities displayed by Nim in the sign-language situation (such as imitation, answering questions, and inserting so-called wild-card signs) were different from the abilities displayed by Lana in the lexigram situation (such as conditional sequence discrimination and variable substitution). Both explanations, however, have emphasized paired-associate learning and the goal-oriented nature of the apes' behavior; the animals were motivated to produce strings of signs or lexigrams in order to obtain a desired object (or event). The approach of these explanations of chimpanzee language behavior may also be useful for understanding some of the abilities involved in human language.

Claudia R. Thompson
Russell M. Church

Department of Psychology,
Brown University,
Providence, Rhode Island 02912

References and Notes

1. D. M. Rumbaugh, Ed., Language and Learning by a Chimpanzee: The LANA Project (Academic Press, New York, 1977).

2. T. V. Gill, in ibid., pp. 225-46; unpublished dissertation, Georgia State University (1977).

3. H. S. Terrace, L. A. Petitto, R. J. Sanders, T. G. Bever, Science 206, 891 (1979).

4. We thank D. M. Rumbaugh for making available to us the complete record of Lana's productions from 29 April 1975 through 29 July 1975, a period that covered the month preceding and the month following research described by Gill (2). The LANA Project during this period was supported by PHS research grant HD-06016. This report was facilitated by research fellowship MH 07741 from the National Institute of Mental Health (to R.M.C.)

* Requests for reprints (and/or a listing of the BASIC program used to simulate this behavior) should be sent to Department of Psychology, Brown University, Providence, R.I. 02912.

Claudia R. Thompson and Russell M. Church. ''An Explanation of the Language of a Chimpanzee,'' Science, 208 (18 April 1980), 313-314

After reading the article carefully, make an informal outline of the points in the article. Read the student outline beginning on page 276. How was it made: How did the student decide to list these points? Is the outline adequate; does it cover the points in the article?

AN INFORMAL OUTLINE

1. Objectives: describe minimum abilities to account for chimp's use of Yerkish keyboard; explain behavior with reinforcement principles (313).
2. Hypothesis: Lana's behavior is result of paired associate learning, conditioned discrimination (313).
3. Lana uses 6 stock sentences (conditioned discrimination) based on cues: whether experimenter has object, whether it is food, whether it is in dispensing machine, etc. (313).
4. Stock sentences contain a name, an activity, an incentive, and may contain other elements—''out-of-room''—(313).
5. Thompson and Church wrote a computer program in BASIC to imitate Lana's decisions (313).
6. Lana needs only yes/no answers to the questions about food to select an appropriate stock sentence (313).
7. Lana can match people, objects with appropriate lexigrams (313).
8. If no food is present, Lana can choose other stock sentences randomly—request food be brought or other incentive, request activity like tickle, groom, etc.—(313).
9. The BASIC program imitates Lana's decisions (313).
10. Thompson and Church had access to and analyzed all Lana's conversations for a 3-month period, based their report on a random sample (314).
11. Results:

    —Lana selects a stock sentence 91% of the time when no experimenter is present (8% errors, 1% nonstock). The program can match this performance (314).

    —when experimenter is present, Lana selects stock sentences 66% of the time (19% errors, 14% nonstock). Thompson and Church feel nonstock sentences represent extension of the system, not departures from it (314).

2

12. Thompson and Church conclude both Lana and Nim were displaying paired-associate learning & goal-oriented behavior. ''The animals were motivated to produce strings of signs or lexigrams in order to obtain a desired object (or event)'' (314).

Depending on your audience and purpose, you might produce different summaries of this article:

A Very Short Summary

Very short summaries are suitable for annotated bibliographies and for reviews of research in which you only wish to mention various articles. These summaries give a short answer to the question, What is it about? For example, here is the abstract published with the Thompson and Church article.

> Abstract. The language behavior of the chimpanzee, Lana, previously described by D. Rumbaugh (1977) can be simulated by a computer model in which the animal selects, depending on context, one of six stock sentences with fixed variable elements.

A Longer, More Detailed Abstract

The longer abstract may be good for a reader who is familiar with the subject, and for most general readers who need a condensed version. The longer abstract attempts to get in as much of the original as possible. It should contain enough information that most readers will feel they don't need to see the original.

> ''An Explanation of the Language of a Chimpanzee'' is a response to the Lana project (D. Rumbaugh, 1977). The authors, Thompson and Church, hypothesize that Lana's ability to use sentences is the result of paired-associate learning and conditioned discrimination. Lana used stock sentences to make requests for food, other objects, and various activities. Using the data compiled by Rumbaugh, the authors wrote a BASIC program that could simulate Lana's behavior. When no experimenter was present, Lana selected an appropriate stock sentence (''Please machine give milk,'' for example) 91% of the time. When an experimenter was present, Lana selected appropriate sentences 66% of the time. The computer program was able to match this performance, and the authors conclude: ''The animals were motivated to produce strings of signs or lexigrams in order to obtain a desired object (or event)'' (314).

ACTIVITY 46 You apply for a job as a research assistant. The job requires you to read magazine and journal articles and write a one-page summary of each article. The researcher needs a concise and accurate summary in order to evaluate what is in the article. To get the job, you must demonstrate your ability to summarize and write clear, correct English. Summarize a magazine or journal article on your research topic. The article you read should be longer than a page, preferably several pages in length. Your summary should be about one page long, double-spaced. Give a full bibliographic note for the article: see p. 190 (magazines) or 191 (journals).

WRITING AN OBJECTIVE ANALYSIS

In a general way, to analyze means to take apart, especially for the purpose of understanding. When we want to understand something, we reduce it to its components to see what makes it tick. If you analyze something physical, like a car, you may produce a long list of parts. Such a list may or may not be very useful to you, depending on your purpose in analyzing. If you want to know how a car is made, you will need the list of parts. But obviously the parts alone aren't enough information. You also need to know what each part is for, and how the parts are connected to each other. Thus, the key to any analysis is the purpose for which you are analyzing. In a given situation you may want to analyze other features of a car such as its styling, its performance, its engineering, and so on. If you want to know not how a car is made but what makes it run, your analysis can ignore everything but the power train and linkage.

Your writing situation will provide a focus for your analysis. That is, during prewriting you must consider why you are analyzing, whether you should analyze for components, organization, process, cause and effect, relationships, functions, behaviors, interpretation, significance, and so on. These factors all depend on your audience and purpose. If you are free to choose your own focus, choose whatever you think will produce the most revealing analysis. Since the purpose of analysis is to illuminate and deepen understanding, choosing the simplest and most obvious features to analyze is usually not a good idea. Anyone can "analyze" a tree into leaves, branches, trunk, and roots—but since anyone can do it, what's the point? You might start with such an obvious analysis while you are brainstorming or mapping out what you already know, but no one's understanding is likely to be increased through that kind of analysis. A predictable analysis is of little use to most adults. You must continue your prewriting until you find a focus that will produce the best analysis for your audience.

An outline is a kind of analysis; it divides a subject into major and minor points and shows how they are related. Assume you have an assignment to analyze the Declaration of Independence. In some situations, an outline of the Declaration might be all the analysis that is required.

Rough Outline of Declaration of Independence

I. Reasons for separation
II. Self-evident truths
    A. Life, liberty, pursuit of happiness
    B. Government by consent of governed

III. Abuses of the governed leading to new government
IV. Many abuses of the king
    A. Abuses, violations of colonial legislatures
    B. Imposition of harmful legislation from "foreign" jurisdiction
V. Colonial petitions to the king ineffective
VI. Petitions to people of England ineffective
VII. Colonies declared independent states

The analysis you produce depends on the analytical strategy you use. Assuming you wish to deepen your understanding of the Declaration, to learn something new about it, you might choose a different strategy for the analysis. You could, for example, do a rhetorical analysis, in which you examine the *ethos*, *pathos*, and *logos* of the Declaration (see Chapter Three). The Declaration is, after all, a message; it is a "rhetorical act." Or, since the document concerns people and their relationships, you could choose to analyze it with the pentad. The pentad treats human use of language as drama, in which there is an act, scene, agent, agency, and purpose. Read the following analysis of the Declaration of Independence. Does the pentad analysis provide any new perspective on the Declaration?

A PENTAD ANALYSIS OF THE DECLARATION OF INDEPENDENCE

The primary ''act'' described in the Declaration is the ''dissolution'' of the political bonds between America and Great Britain. From the British point of view, it was a rebellion. Other acts in the document refer to the many acts of oppression by the king. But the most interesting act may be the ''act'' of writing the Declaration itself. To the British it was the insolent and irrelevant dare of a band of outlaws; but to the Americans it was a birthday announcement to the world at large—the birth of a new nation.

Description of Declaration as an Act, various Acts within

The ''scene'' of this act was first of all Philadelphia, 1776. But the larger scene was continental America and eighteenth-century Great Britain, the one a struggling colony of 13 states, the other the greatest power in the world at the time. By 1776 the Revolutionary War was already a year old, and England had sent a large army to put down this ''rebellion.'' Thus, the colonists were already under siege when they put forth their declaration. Other scenes in the act include the colonial Seas, Coasts, and Towns which had been ''plundered,'' ''ravaged,'' and ''burnt'' by the king.

Description of scenes in Declaration

Use of supplementary information, historical background

The ''agents'' in the Declaration are the King of England, described as an abuser, usurper, and oppressor, whose goal is ''absolute Despotism'' and ''absolute Tyranny.'' The Declaration says of the king, ''A Prince, whose Character is thus marked by every act which may define a Tyrant, is unfit to be the Ruler of a free People.'' And, in fact, he is not really king over the colonists, for, ''He has abdicated Government here, by declaring us out of his Protection and waging War against us.'' On the American side, the agents are the elected Representatives of the 13 states. These are lawfully appointed and lawfully acting men, who have a ''decent Respect to the Opinions of Mankind.'' They obey the ''Laws of Nature and Nature's God.'' They are men who would not change

Description of Agents in Declaration

Use of quotation for substantiation

Analysis of rhetorical strategy: bad agents versus good

governments for ''light and transient'' causes, who would rather suffer evils as long as they can. They have been ''patient'' sufferers. All these injuries the colonists have protested ''in the most humble Terms.'' Then too, they see themselves as the ''Brethren'' of British citizens in England. The Americans are, in the words of the Declaration, ''the good People of these Colonies.''

Description of Agencies in Declaration

Illustration of rhetorical strategy: loaded language

The agencies used by the agents are several. The king's most impressive agency is of course the British army. In addition to the regular army, the king uses ''large armies of foreign Mercenaries to compleat the works of Death, Desolation, and Tyranny, already begun with circumstances of cruelty and Perfidy scarcely paralleled in the most barbarous Ages, and totally unworthy the Head of a civilized Nation.'' As if that weren't enough, the king is also aided by ''the merceiless Indian savages, whose known Rule of Warfare is an undistinguished Destruction of all Ages, Sexes, and Conditions.'' Overall, the king uses the power of the British Crown to frustrate and torment the colonists.

Agencies of morality, reason, argumentation contrast with force

Declaration is its own Agency—a legal document

Declaration is an argument for the court of public opinion

The colonists, on the other hand, rely on the moral superiority they get from Nature's God, ''Divine Providence.'' Whereas the king always uses force, the colonists use reason and argument. They appeal to morality and logic. The Declaration itself is an agency of change; the colonists have produced a legal document. The king responds to no authority but his own; the colonists respond to the sense of justice that governs all human societies. The Declaration is an argument or a legal brief. It sets forth the laws that are involved, especially the ''right of the People to alter or abolish'' abusive governments; it ''proves'' the many abuses of the king, including his ''abdication'' of government in the colonies; and it concludes that ''We therefore . . . declare . . . that these United Colonies . . . are absolved from all Allegiance to the British Crown.''

3

Finally, the purpose behind the Declaration is stated as to ''declare the causes which impel [the colonists] to the Separation.'' But the document isn't addressed to the king. There was no need to ''declare'' themselves separated from England, since the king was already at war with them, and the king would have been unimpressed with such a message anyway. Instead, the given purpose is ''a decent respect to the Opinions of Mankind.'' There were two audiences for this message: one was the other nations of the world, particularly France, on whom the new nation would have to rely for aid to survive. The other audience was the colonists themselves. Though they were at war with the Crown, the idea of revolution and breaking away from England entirely was not popular with everyone. It was necessary to convince the colonists of the severity of their oppression and the morality of their action. The Declaration was both a call to arms and a plea for help.

Transitional phrase

Purposes of Declaration analyzed

Purpose related to audience

The Process of Writing an Analysis

Look for an article you can analyze, preferably one on your research topic. The more knowledgeable you are on the subject, the more credible your analysis will be. Avoid articles that are too short or too simple to analyze. On the whole, your analysis will be more worthwhile if you use a longer, more demanding article.

Decide what kind of focus you will use. What will best explain the article—a plain description of major and minor points? A dramatistic analysis like the pentad? Something else? You must consider your writing situation: Who is the audience, what is the purpose of your analysis? (See Reading for Analysis, Chapter Three.)

Make a rough outline that incorporates your focus. It isn't enough to mark the article with a highlighter. Making your own outline will force you to synthesize and assimilate the material. Your outline can be informal, but it should have enough form that someone else can read it and follow your thinking. You may be required to turn in a formal outline with your analysis (ask your instructor).

Write a rough draft following your outline, but be alert to possibilities for changes not covered by your outline. Make sure each point you make is well substantiated with examples from the article you are analyzing. Quote from the article to illustrate your points.

Review your draft. Does it cover the article well? Are there other points you could make? Do you have enough examples? Revise by adding examples and additional points of analysis; cut loosely worded sentences and any information that doesn't contribute to the analysis.

Polish your writing. A good analysis can be lost in poorly written sentences. Proofread for errors. In the end, your analysis should be a readable composition, even for those who haven't read the article you are analyzing.

ACTIVITY 47 You are a committee member in local government. An article or report of interest to your community has come to your attention. You are elected to write an analysis of the report for the committee and possibly for release to the community at large. Select a suitable magazine or journal article, and prewrite and revise your analysis; then write a detailed analysis several pages long.

WRITING A CRITIQUE

A critique is a kind of analysis, but being "critical" implies something more than merely analyzing. Criticizing implies making judgments, expressing opinions. Specifically, a critique is an evaluation. In an analysis we ask What are the components? How is it put together? but in a critique we ask What is its value? To evaluate anything implies judging it. Much depends on your writing situation. If you are evaluating for your own use, you can be guided by your own subjective opinions. But if you are evaluating for someone else or your critique will be published or your evaluation is a component of a research paper (your conclusions in the paper are based on your evaluations), you must make sure you are well read in the subject. You must know the subject well enough to make judgments about the quality of the information. Instead of relying on your own subjective opinions or feelings, you must use a set of criteria that others could also use (see Chapter Three, p. 83). Your judgments must be verifiable. That is, using the same criteria, others should be able to agree with your judgments, or at least understand how you arrived at them. Thus, in a critique, you must not only give your judgment, you must say why. (See Substantiation, p. 8.)

The primary requirement for any critique is that you know what you are talking about. You needn't be the world's greatest authority, but obviously the more you know about Shakespeare, the more credible your critique of *Hamlet* will be. The more knowledgeable you are on any subject, the more you are likely to know the appropriate criteria to use in a critique. That is, in addition to general criteria that might apply to almost any writing situation, there are more specific criteria that apply to specific subjects. Read the following example of a critique. What criteria has the author used? How can you tell that Denise is knowledgeable about this subject?

1

TEACHING LANGUAGE TO AN APE
DENISE TAYLOR

Introduction identifies subject of critique.

''Teaching Language to an Ape'' by Ann and David Premack is a report of their experiments with the chimpanzee Sarah. The Premacks describe Sarah's plastic–token language and how they taught her to use it. She had 130 terms and used them with 75 to 80% reliability. Her tokens were arbitrary shapes and colors. With them Sarah could name objects, make requests, answer questions, and classify by size, shape, and color. The Premacks conclude that Sarah has learned a simple language, similar to that of a two–year–old child's.

Introductory strategy: summarize article.

Criticism of amount of detail, authors' use of general language.

The article is a summary of six years of experiments. Therefore there is little detail in the article, and it is hard to judge the validity of anything in it. For example, we are told it took ''several'' trials to get Sarah to learn something (95) or that she made ''many'' errors (96), but never how many. Or we are told that Sarah had a 75 to 80% reliability score (92), but there are no data on numbers of tests or controls on the tests. The article reports some of the interesting findings for a general audience.

Criticism of inexact, incomplete data.

Judgment that article is aimed at general audience

The source of the information is the authors' firsthand observations. They give a brief mention of prior research (92) but do not document their sources. In their conclusion they speak of ''linguists and others'' (99) who are skeptical of ape language abilities . . . without identifying who they are. Their article seems somewhat informal for a scientific report of research.

Criticism of authors' use of sources.

Judgment that article is informal.

The authors' chief assumption is that what Sarah was doing was using a kind of language. Since they never mention any other possibility, it apparently does not occur to them that Sarah might be simply finding some way to select the right tokens—like unintentional cuing from the trainers. Further, they assume that Sarah's behaviors correspond to the behaviors they wanted her to have. When Sarah <u>selects</u> the token for

Analysis of authors' assumptions.

''color,'' the Premacks say she has <u>classified</u> the red objects (96-97). Or when she <u>selects</u> the token for ''same'' they say she has <u>named the relationship</u> between objects (95). But from Sarah's point of view all her tests are the same—Sarah is required to select the right token for her reward. The authors' assumption that her behavior qualifies as language use may be too strong.

Criticism that authors use *a priori* reasoning.

The Premacks' use of examples is limited. Their examples tend to be general and lacking in detail. Often we are told only that Sarah picked the right token. Their conclusion, for example, that Sarah made mistakes in a certain test because she was ''expressing a preference'' for one kind of fruit over another seems a large leap in logic (96). This test gives her a different fruit reward for naming the fruit in front of her (in previous tests the fruit in front of her was always the reward). We are told Sarah had a hard time learning this: ''Sarah take banana if-then Mary give chocolate.'' A similar kind of reasoning may be observed in their discussion of the ''compound sentence.'' The authors believe Sarah can grammatically analyze the sentence and understand what object goes with what receptacle by the rules of grammar (99). But Sarah learns this compound sentence as the last of three steps:

Criticism that authors are too general, lack detail.

> Sarah insert apple pail. Sarah insert banana dish.
> Sarah insert apple pail Sarah insert banana dish.
> Sarah insert apple pail banana dish. (98)

Use of documentation, quotation for substantiation.

A simpler conclusion might be that Sarah continues to recognize what she must do even after minor parts of the sentences have been edited out. It appears—and some researchers have said—that the Premacks are ignoring Occam's razor, ignoring simpler explanations.

Judgment that authors ignore Occam's razor.

The Premacks are important researchers in this field. They have published books and several reports and articles, and their work is frequently

Concessions: acknowledges authors' expertise, value of their study.

cited by other researchers. Their article represents a major development in the ape language research. Though different from the Washoe or Lucy experiments, it is one more confirmation of the idea that apes may be capable of some kinds of language behavior when provided a language system they can use. From that perspective, it is important to both scientists and the general public. Despite whatever problems there may be in the techniques, if this research is correct—that apes may in fact be able to use certain kinds of language—old ideas about the difference between humans and apes, and possibly other animals, will have to be reevaluated.

Discussion of authors' style. The article is readable, contains little jargon.

Fortunately, the authors are very readable. Though they are reporting scientific experiments, the subject is nontechnical. There is very little jargon other than a few terms from psychology like cognitive, perceptual, displacement, and so on. Even these are not hard to understand in context. The authors' sentences are relatively long, but the language in them is simple enough that they are not hard to read:

Illustration: long sentences are readable due to simple language.

> On the basis of these demonstrated conceptual abilities we made the assumption that the chimpanzee could be taught not only the names of specific members of a class but also the names for the classes themselves. (95)

Description of tone: mostly objective, scientific.

The tone of the article is mostly objective, scientific reporting. But here and there the authors have used less objective language: ''The chimpanzee, which was then about five years old, was allowed to eat the tasty morsel while the trainer looked on affectionately'' (95). Then too, giving the chimpanzees names helps to humanize them. And reporting the simple concrete details of some of the tests (Sarah gets slices of banana and apple and chocolate) reduces the abstract sound of research. Telling stories appeals to readers:

Illustration of lapse into emotional language, less objective tone.

> Sarah learned the names of some of the recipients the hard way. Once she wrote

4

''Give apple Gussie,'' and the trainer promptly gave the apple to another chimpanzee named Gussie. Sarah never repeated the sentence. (95)

Illustration of authors' use of anecdote.

Furthermore, there are pages of illustrations of the plastic tokens as well as illustrations (sketches) throughout the article of Sarah and her sentences. These illustrations are visually appealing as well as very helpful in understanding the concepts. So while the overall tone is professional and objective, the authors have made an effort to make it appealing to the reader. The article has a lighter tone than many scientific articles.

Description of other efforts by authors to appeal to readers.

The authors' diction is excellent. Overall the words are simple; there is nothing in the article an educated reader is not likely to know. The authors use metaphoric language, referring to ''brain mechanisms'' and ''cognitive mechanisms,'' (95) but their metaphors are not poetic. The article is written mostly in the third person. Rarely, the authors call themselves ''we'': ''In teaching Sarah we first mapped . . .'' (95). Many verbs were passive: ''The interrogative was introduced with the help of the concepts 'same' and 'different''' (95). Thus while the authors are obviously well educated, the goal of their diction is simplicity, and at that they have been successful.

Critique of authors' diction.

Illustration of authors' use of metaphor, first person, passive verbs.

Though there is an occasional longer sentence, the authors' sentences are relatively short, polished, and highly readable:

It is not necessary for the names to be vocal. They can just as well be based on gestures, written letters, or colored stones. The important thing is to shape the language to fit the information-processing capacities of the chimpanzee. (95)

The Premacks use rhetorical questions: ''Why try to teach human language to an ape?'' (92). ''What form does Sarah's supposed internal representation

Illustration of authors' short sentences.

Criticism that there are no memorable sentences.

take?'' (98). While there are no bad sentences, and on the whole they all seem readable and polished, there is no especially memorable sentence. This style neither attempts nor achieves anything other than the transfer of information. The sentences are competent academic writing, nothing more.

Judgment that the writing is competent.

I think the style is good, all things considered. I believe the authors were forced to deal with the limits of writing for a popular audience, and they were able to do so without condescending. The style is an appropriate one for the subject and audience.

Conclusion that the style is good.

Overall it is a good article. On its own it is interesting popular science, good for a few minutes of relaxed reading. In the easy-to-read, polished style of the Premacks, the article will provide dinner-table conversation about science. But it is also important for the ape-language research. As one of the early reports introducing a new mode of communication for chimps and confirming their ability to use language, the article should prove useful to other researchers.

Overall judgment that the article is good: informative for general readers, important for researchers.

Premack, Ann James, and David Premack. ''Teaching Language to an Ape.'' Scientific American 227.4 (1972): 92-99.

The Process of Writing a Critique

Writing a critique puts you in a position of superiority over the article you are evaluating. Therefore it is important that you be scrupulously fair and absolutely accurate. You must read the article very carefully as many times as necessary to make sure you understand it clearly. If necessary, read other articles that will help you to understand the subject.

Make a careful outline. It is not enough to have a general understanding of the article. You must have an exact understanding—your own credibility is at stake.

Consider your writing situation: your audience and purpose. What point of view will you use—disinterested, impartial reviewer? Student, learner, someone who appreciates the article? Vengeful critic ridiculing the ignoramus who wrote the article? Your own point of view will influence the criteria you select and the tone of your critique. The *appropriate* point of view depends on your situation.

Make a rough draft or rough outline for a draft using the criteria on style and content in Chapter Three. For your first draft your goal should be thoroughness and accuracy. Try to use all the criteria. For each point you make, you must cite examples from the article: quote or paraphrase illustrative material. Each paragraph you write must cover one and only one point.

Review your draft. Decide which paragraphs have the most information; cover the most important criteria. Delete anything you think doesn't really contribute to the critique.

Revise your draft. Find additional examples. Consider the organization of your paragraphs. In general, order of importance is the preferred order, but this is not an inflexible rule. You must imagine the effect on your reader of different organizational patterns. Should the most important material come first—or last? (See Chapter Six.)

Polish your writing. Revise your sentences for flow and emphasis. Look for repetition; too many sentences that sound similar will have a bad effect on your readers. Proofread carefully.

ACTIVITY 48 You are the reviewer of an educational journal publishing articles of interest to teachers and the educated public. Find a journal article on your research topic, an article long enough and challenging enough for a review. Using the criteria in Chapter Three (p. 83), write a critique of the article's style and content.

WRITING A COMPARISON

One very good way to understand anything is to compare it with something similar. One moon rock is simply a mystery, but two moon rocks are a wealth of information: each provides a reference for the other. By comparing the new or strange with the old and familiar we increase our understanding; one moon rock and one earth rock are the beginning of wisdom. When we compare, we point out similarities and differences; it is not necessary to say "compare *and* contrast." Contrast is inherent in a comparison, and it is often the contrasts that are most informative.

The point of comparison is to discover something new. A predictable comparison is no more interesting than any other predictable analysis. And as in analysis, a comparison becomes credible and worthwhile in relation to your expertise in the subject. You need criteria on which to base a comparison, generally the same criteria you would use to analyze a single object or idea. Your comparison should be substantiated and verified. (See Chapter Three, p. 83–86.)

Read the following comparison. What criteria has Jill used? Is her comparison credible, reasonable? Does she sound fair to both authors?

AN EMOTIONAL AND LEGAL FOCUS ON PRAYER
IN THE SCHOOLS
JILL GREGOR

The controversial issue regarding religion in public schools has created an American dilemma. Even though there have been Supreme Court rulings, the lower courts continue to wrestle with the definition of terms. Because of the growing confusion concerning prayer in school, several informed writers are communicating their views. Two such articles discuss this dilemma; one appears in The Humanist and the other in The American School Board Journal.

Introduction identifies subject.

''To Pray or Not to Pray: Requiring Students to Reveal Their Religious Preferences'' was written by Haig Bosmajian and published in The Humanist, a professional journal. According to Dr. Bosmajian, ''Freedom of speech is a right guaranteed by the Constitution, but what about the right not to speak?'' While freedom of speech has always been a characteristic of free societies, versus restriction of speech in totalitarian societies, the right not to speak has been overlooked and not associated with our constitution. By contrasting a totalitarian nation with a free society, Bosmajian clarifies that the constitution protects the right not to speak. Bosmajian explains that ''another crucial difference is that in the totalitarian nation citizens are and can be denied the right not to speak, may be denied the right to remain silent, whereas in the free society the state cannot compel, cannot coerce the citizen to profess his or her political or religious preferences and beliefs.'' Citing several court cases, he indicates a concern that Americans are not aware that remaining silent is also a constitutional right. The focus of his article is to address the possibility that the rights of one group can be infringed upon by another group. He refers to ''the removal of the nonparticipating children from the

Paragraph introduces, summarizes first article.

Identifies author's technique.

Refers to author by name.

Use of quotation to illustrate.

classroom'' as an infringement on their rights as they are not allowed to remain silently seated in the classroom. He appeals to an uninformed yet freedom—conscious reader. Consistently, Dr. Bosmajian's purpose is to make the reader aware that the right not to speak is constitutional.

Identifies author's purpose.

Introduces, summarizes second article.

The article ''Learn to Distinguish Teaching from Preaching'' by Dr. Benjamin Sendor is published in The American School Board Journal. Being an attorney as well as a professor of the school of law, the author attempts to inform the reader concerning the legal issue of prayer in schools. He describes the legal steps and concerns of a judge when dealing with this controversial issue. He states, ''Teaching about religion is constitutional, but advancing specific religions is not.''

Contrasts the authors' points of view.

Both authors identify the problem of what constitutes prayer in school but describe the circumstances from different perspectives. Dr. Bosmajian feels it is important to relate to the reactions of the students as he entitles his article ''To Pray or Not to Pray: Requiring Students to Reveal Their Religious Preferences.'' He states, ''Students who are required to leave the classroom because they cannot in good conscience participate in the prayer ceremony are clearly identified as the unorthodox.'' On the other hand, Dr. Sendor emphasizes that there should be a legal understanding of terms in his title ''Learn to Distinguish Teaching from Preaching,'' which requires the defining of these two words. In the *Crockett* v. *Sorenson* case, Dr. Sendor refers to the Federal District Judge Jackson L. Kiser as ''giving school board members and school administrators a lesson in separating teaching from preaching.''

Analysis of first author's style.

The style of writing used by each author is different, a result of their appeal to different audiences. Dr. Bosmajian's article in The Humanist communicates to the emotional reader. His vocabulary can be easily understood as he puts each court case in story form, in which the reader can identify with the main character, a child. When speaking of ten-year-old Terry McCollum and how he was ordered out of the room for not participating in religious instruction, Bosmajian speaks of Terry sitting ''at a desk in the hallway where students passing by teased him, thinking that he was being punished.'' The reader forms a bond with the child and shares his agony.

Illustration of author's simple language, use of anecdote.

A biased viewpoint exists throughout Dr. Bosmajian's entire article. He encourages the reader to disapprove of religion in schools at the very beginning of his article when he states, ''The state compelled the students to reveal their religious beliefs and inclinations; the child's religious preference, an otherwise private matter, had been made public.'' He definitely shows his bias as he says, ''School children not participating in the prayers or the Bible readings, who are asked or required to leave the classroom, are as a result separated both physically and spiritually from their classmates and their teachers.'' Dr. Bosmajian's article attempts to persuade the reader to think as the author does.

Analysis of author's bias.

Reveals author's argumentative, persuasive purpose.

However, Dr. Sendor seeks to present the facts objectively, allowing the reader to think independently. He refers to court cases without emotion by stating only the facts. He lists guidelines suggested by the court to distinguish teaching from preaching. These guidelines are presented in order to establish an organized approach for school systems wishing to offer religious courses. He also lists questions having to do with the constitutionality of prayer in schools. By including legal guidelines and questions to determine what is religion in schools, the author encourages readers to use these methods to form their own opinions.

Contrasts second author's objective treatment.

The cases mentioned by Dr. Bosmajian are emotionally presented for the purpose of persuasion. He does not supply facts which would enable the reader to analyze the case in depth. His article is much longer than Dr. Sendor's because he

Criticizes author's use of emotion, opinion.

continually expresses his own opinion. Dr. Bosmajian states, ''However, with the public school prayer comes the state's command: either pray or leave the classroom; pray or stand in the hall.''

Illustrates with quotation second author's objective approach.

Dr. Sandor gives a detailed account of each court case used in his article. He simply says, ''The class routine included Bible teaching, prayer, and hymn singing until 1982, when the prayers and hymns were dropped, leaving a curriculum of Bible teaching and nondevotional music.'' This nonpersuasive approach gives the reader a view of the legal issues in the dilemma of prayers in school.

Summarizes authors' similarities, contrasts their purposes.

Both authors address the issue of prayer in the schools. Each uses court cases as background material, yet their intent regarding the reader is different. Dr. Bosmajian's purpose is to persuade the reader to empathize with the child who chooses not to pray in the school. Dr. Sendor provides thought-provoking but unbiased information allowing the reader to independently interpret the evidence concerning school prayer. Though opposites in purpose, Dr. Bosmajian and Dr. Sendor exhibit a consistency in style, accomplish their goals, and provide well-written articles.

Concludes both articles are well written, achieve their goals.

Works Cited

Bosmajian, Haig. ''To Pray or Not to Pray: Requiring Students to Reveal Their Religious Preferences.'' The Humanist Jan. 1985: 13-17.

Sendor, Benjamin. ''Learn to Distinguish Teaching from Preaching.'' The American School Board Journal 17 Apr. 1984: 6.

The Process of Writing a Comparison

Find two magazine or journal articles on the same subject. (Newspaper articles are usually too short and not good for this assignment.) The two articles need not be identical. One article on AIDS from a scientific point of view and another on AIDS from a human point of view can provide the basis for a comparison. If you have been collecting articles on your research topic, any two of them can be used. Otherwise, use the library indexes to find articles on a subject that interests you. Avoid articles of very different lengths. If one article is much shorter than the other, you will not have enough material for a balanced comparison.

Read both articles carefully. Outline both so that you will have an exact understanding of the points they cover.

Select the criteria you will use (See Chapter Three). Make a rough outline of the points you think are worth making and the supporting material from the articles; quote and paraphrase examples to illustrate each point. The more criteria and supportive material you use, the more thorough your analysis will be.

You need an introductory paragraph to tell the reader what you are doing. You also need a brief summary of each article for the benefit of readers who have not read them. (See Jill's comparison, p. 293.)

Organize your comparison for effective presentation of the points. You have three choices:

1. *Comparison of the wholes,* in which you first analyze one entire article, covering all the points of analysis you intend to make, and then move on to the second article and cover all its points of analysis. This is the simplest plan and is preferred by many students; however, many readers dislike it because it requires them to remember what was said about the first article while reading about the second.
2. *Comparison of the parts,* in which you discuss both articles together, point by point. For example, you might have a paragraph on the tone of articles 1 and 2, a paragraph on the assumptions of 1 and 2, a paragraph on the diction in 1 and 2, and so on. This is a more sophisticated organization, one that requires more deliberate planning on your part.
3. *Comparing a mixture of parts and wholes.* You cannot really mix the two plans, but you can give that appearance by sometimes discussing both articles in a single paragraph, sometimes devoting separate paragraphs to each article. Mixing the two methods in this way allows you to overcome the monotony of repeating the same pattern throughout your paper.

Remember that you must write a readable composition; in effect, your comparison is an essay. After you have written a first draft, let it "cool off" for a day or two. Revise by supplying more criteria, more points of analysis, more examples. You must be especially critical of your paragraphs in a comparison; make sure each paragraph contains only one point. Especially avoid writing one large paragraph about style; write separate paragraphs about readability, sentence structure, diction, and so on. It's a good idea to get feedback on your final draft before submitting it for a grade.

ACTIVITY 49 A controversy has arisen over two articles appearing in magazines or journals. The articles cover the same event or subject, but differences or similarities in them have raised questions about objectivity, accuracy, and professional ethics. As an expert on nonfiction writing, your job is to write a careful, objective comparison of the two articles. Your analysis will serve as the basis on which the committee will attempt to understand the two articles' similarities and differences and to answer questions about them. Select two articles on your research topic. Your analysis may be three to five pages long. You must document what you say about the articles.

WRITING A REVIEW

A review of a book or a performance like a concert, film, or play can be either an objective analysis or a critique. A review can be a single sentence or several pages long. When a work first comes out, "reviews" are often simply descriptions:

> Lloyd Nielsen's new book Monkey'n Around is a personal narrative of the lives of several research animals at the Institute for National Research. Part history and part argument, the book details the indignities and cruelties committed on lab animals in the name of research.

This kind of "review" can be called objective since the reviewer does not express opinions but merely tells readers what they will find if they read the book. Such reviews are useful to readers who want to know what is in a book without the distortions of a reviewer's personal biases. In essence, the review is simply an extended analysis and requires little more than accuracy from the reviewer.

However, much more familiar to most people is the critical review, the critique. "Critical" need not imply negative—a critique can be entirely positive. If you think a book is excellent and has no flaws or weaknesses, you need not say anything negative about it. "Critique" means evaluating, expressing judgments. But because critiques are often negative, the word has come to have a negative connotation that it does not deserve. In any case, to write a critique of a book means to evaluate it, to say whether it is good or bad. To write a balanced critique you should point out both strengths and weaknesses in the book. Like the review of an article, the review of a book or performance requires that you use criteria and cite evidence from the work. (See Chapter 3, Critical Criteria.)

Read the following book review. What criteria has Sharon used? How credible is Sharon as a reviewer (do you believe her criticisms are valid?) and what is her credibility based on? Has she been fair to the book?

STUTTERING
SHARON ATTAR

Introductory strategy: historical allusion.

Stuttering, the repetition and prolongation of sounds, is a disorder which has plagued man since biblical times, and continues to baffle speech pathologists, psychologists, and neurologists, as well as concerned parents and other laypeople as to its etiology and remediation. Consequently there have been many books, articles, and pamphlets written by various authors explaining—or attempting to explain—the mystery of stuttering.

Introduction identifies subject.

Stuttering by Dr. Edward Conture (Englewood Cliffs: Prentice-Hall, 1982) is a book intended for the graduate-level student speech clinician as well as practicing speech-language pathologists, and discusses the author's impressions concerning the remediation of stuttering.

Criticism of style, figurative language.

As one begins reading Stuttering, one thing that is quickly apparent is the author's use of figurative language. Often this use of description makes for clearer and more interesting reading, but here there is too much of a good thing. For example, the author says:

Illustration of author's use of figurative language, analogy.

> Just one swallow does not make a summer, neither does one instance of stuttering mean a person is a stutterer. Individuals who have just experienced one bout of alcoholic intoxification [sic] do not immediately seek out their local AA chapter; likewise, neither should a temporary period of within-word disfluences mean that people should seek out therapy for stuttering. (15)

He also compares stuttering to love (139) and falling off a cliff (149) and remarks:

Illustration of author's use of exaggerated analogy.

> Much like the medical specialist (allergist), whom the current writer sees, and who treats him for allergies, the allergin dosage used in the desensitization injections we receive is slowly increased but leveled off at

> any signs that the desensitization shot is precipitating an allergic reaction (sneezing, wheezing, or itching). If need be, the allergist may even back down the level of our allergin dosage, say from 0.3 cc to 0.2 cc, if the higher level causes us too many problems. Likewise, the speech and language pathologist is not trying to elicit stuttering but to slowly and systematically increase the child's ability to react fluently to communicative situations which are increasingly stressful. (75)

The author's use of such lengthy, complicated analogies to illustrate relatively simple concepts only serves to confuse the reader.

In conjunction with the author's overuse of figurative examples, his language seems on the whole to be rather overblown, stressing information and terms that seem easy enough to understand without extra emphasis. He comments:

Illustration of author's overblown writing, unnecessary explanations.

> The word apt is purposely used because the relative definition involves a statement of probability rather than certainty. For example, comics who attempt to imitate stutterers as part of their act are apt or most likely to select within-word rather than between-word disfluences. The phrase most likely denotes an event that is relatively (probably) rather than absolutely (certainly) likely to occur. (8)

Words and phrases like ''apt,'' ''most likely,'' ''relatively,'' and ''absolutely'' are pretty straightforward and do not seem to warrant the extra explanation the author has provided.

This oversimplification makes the author seem almost condescending. He also uses the ''we'' form very often throughout, making his words sound like he is trying to place the weight of all professionals behind him. He says, ''We present concepts, ideas, notions, procedures . . . that we

Judgment that author is condescending.

Criticism of author's use of "we" for authority.

believe pertinent,'' and ''We think, however, that this book provides both the student-clinician and the practicing speech-language pathologist with meaningful insights . . .'' (2). In fact, his attitude seems to be one of the leader of a secret club initiating the plebes. Laypeople, he implies, especially parents, haven't a clue as to raising and spending time with children without continually persecuting them for real (or even imagined) disfluences, although he comments:

Illustrates author's concession to parents.

> Many of us will eventually become or already are parents, and we should recognize that parents encounter many of the same problems that we speech and language pathologists, as people, encounter . . . we should not be willing to cast the first stone at parents who, after all, are people like the rest of us, with all our human foibles and fortes. (64)

This is a noble thought, but he refutes it with negative comments like, ''We, as speech-language pathologists, should not pester the child the same way the parents and the other children may be pestering him or her'' (56), and ''This is an obvious negative evaluation of the teen's personality and is also probably harassing the client in the same way that his or her parents do'' (103). His comment concerning the ''typical'' parent is especially insulting, both in content and language.

Refutes author's concession to parents by quoting representative negative comments.

> We remember one mother who could listen all day to her son's prolonged sounds and not bat an eyelash; however, let that same son repeat the same sounds in company and the mother would respond with, ''You know better than that,'' or ''Stop that this instant.'' (23)

Quotes an especially revealing illustration of author's attitude.

Besides some difficulties with language, Dr. Conture has a problem of getting off on tangents that have little to do with his subject and only serve to detract from the other information. For

Criticism of book's organization, unity, coherence.

example, in a paragraph that is supposed to deal with ''objectively changing speech behavior'' (96) of teenagers, there is suddenly a large section dealing with the use of a VU meter, a section that probably would have been better placed in the chapter devoted to equipment use in the clinical setting. This tendency to get off the track is also seen a few pages later in a section dealing with ''practice and/or carryover of change'' (99–101) where he gets caught up in a heated discussion of why phone companies should ''develop special, low-budget phone services for the speech-language pathologist'' (101), which really has little bearing on the subject he is presenting.

Other examples, while not out of context necessarily, seem contradictory and make little sense. For example, Dr. Conture begins one paragraph by saying, ''Marital disharmony, in particular, comes into consideration with parents of stutterers,'' but refutes it by ending with the observation that ''our clinical observations suggest that the divorce rate among the parents of stutterers is no greater than that of the average public; perhaps the parents of stutterers are more likely to remain together . . .'' (36). This type of _non sequitur_ comment is also seen in another instance when he is discussing ''questioning the client and associates regarding the _who_, _what_, _where_, _when_, and _why_ of the stuttering problem. . . . If the clinician wants black-and-white, yes-and-no responses from the client, then questions can be structured accordingly ('Are you a boy?')'' (23). _This_ writer fails to understand how the answer to a question such as ''Are you a boy?'' could answer questions about the ''who, what, where, when, and why'' of stuttering or any other problem. Dr. Conture also asks later, concerning a child with both a nasal-quality voice and a stuttering problem: ''Which came first: the stuttering as reaction to the nasal voice quality or vice-versa. . . . Is there any relation between the stuttering and the nasal voice quality'' (26)?

Criticism of author's reasoning, coherence.

Identifies author's use of logical fallacy.

Evaluates the quality of author's question.

The author fails to show that this is a significant question or that if he knew the answer to his rhetorical question he would be any better able to treat the problems of his client.

Evaluates amount and quality of author's evidence.

Obviously the writer's attitudes, and indeed his style, reflect the choice of information in his book. Most of the book seems to be the result of the writer's own observations and practice as a speech pathologist, and most of it is sketchy, with almost no actual statistics. The writer seems to be speaking in generalities much of the time. He says rather defensively, ''Now before you start saying, 'Where are the data?' let me reiterate that I said <u>some</u>, not all.''

Illustrates with quotation author's lack of data.

Throughout his work, however, Dr. Conture makes many references to other authors, but these references seem very often to be irrelevant and distracting rather than enlightening. For example, in one section (3) he states, ''We are not referring to behavioral approaches restricted to this or that school of learning psychology (see Reynolds, 1968; LeFrancois, 1972; Hill, 1977 for overviews of learning psychology)'' and later (10) that 'these behaviors appear to become reinforced in much the same way that the green shirt the pitcher wore on the day he pitched a no-hitter takes on special properties (see Skinner 1953 for discussion of superstitious behavior).'' Neither of these references—essentially rather ''show-offy'' content notes—is very relevant to the topic of stuttering, even to the young clinician.

Criticism of author's use of sources.

On one occasion when Dr. Conture does discuss another author's work, he barely skims its surface and does not elaborate on the other author's opposing view, saying, ''Van Riper's (1973) cautions in this area are well taken; however, it is my experience that with patience and persistence on the part of the speech-language pathologist, parents can learn, and even come to enjoy, reading to their children'' (53). He does not, in effect, give Van Riper a chance to ''speak.''

Criticism of author's unfair use of source.

6

Although some of his analogies are difficult to follow, Dr. Conture does give a few excellent ideas for use when helping stutterers understand some basic concepts of speech and stuttering and how they work. His comparison of a garden hose to the speech process in teaching a small child (56-57), and the comparison of touching the thumb tip to the fingertips with speech seem both interesting and relevant. Comparisons such as these are quite good, but unfortunately they are rather overshadowed by the more negative aspects of the book. They are an example of good information placed in a not-quite-good-enough surrounding. For a graduate-level textbook, this seems both too simplistic and too overwritten to appeal to anyone with some background in speech pathology.

Concludes author has difficulty with style but has a few good ideas.

Overall judgment that book is too simplistic, overwritten.

The Process of Writing A Review

In prewriting your review, one way to start is simply to brainstorm anything and everything you can remember about the book. You can also try a cognitive map of your ideas about the book. The idea is to look for what sticks with you about the book, what impressed you. When you have exhausted your prewriting efforts, you should begin to look over your list of criteria. What criteria match with items in your brainstorming or mapping? From this you can begin to construct a very rough, preliminary outline for your paper.

It is always a good idea to start with a brief summary of any book, play, or film you are reviewing. You should not assume the reader has read or viewed the performance. Even if you are certain the reader is familiar with the performance, you should not write your review with that assumption. Write your review so that it can be read by anyone, and that means supplying a brief summary. At the very least your reader needs a few identifying sentences to know you are discussing a book instead of a play.

You need not comment on everything in the book, but the more items of analysis you have, the more thorough your review will seem to your readers. Remember to discuss only one point per paragraph. Give details from the book to illustrate your point—the more details the better. You should assume that your review is an argument and that you need to verify what you say by citing examples. It is too easy to say the author's language is appealing or dull; such generalizations are mere opinions, after all. You must quote the author's language, explaining why you have such an opinion. And usually the more examples you provide, the better. If you cite only a single example of anything it looks like the point is not worth making or is not typical of the work.

You can arrange your paragraphs in any order that makes sense to you, but order of importance is common, progressing from least to most important (sometimes called climactic order). (See Chapter Six.)

ACTIVITY 50 A new book has come out on your research topic (you are an expert). You believe the book deserves a review because it is very good or very bad or some other reason and you undertake to write the review for a scholarly journal. Find the guidelines for submissions in the front of a scholarly journal in your field. Select one of the books from the subject you are researching for your research paper, and write an in-depth review several pages long.

WRITING A REPORT

A report is usually a compilation of information. Situations in which we "just want the facts" are ideal for reports. To write a report you must act as a reporter; like a newspaper reporter, you must observe events, collect information, and then reconstruct—usually in chronological order—the "news." To write a report from secondary sources requires you to collect and read everything you can find on the topic and then organize the information for accurate and efficient transmission. Of course, everything depends on your situation, your audience and purpose, but an objective report usually calls for compiled information without opinions or biases from you.

Read the following student report on cocaine trafficking in Latin America. Has the student been thorough? Is he credible (are you able to believe that his report covers the cocaine trafficking problem)? Note that Glenn has used the name-and-date style of documentation.

FIGHTING COCAINE TRAFFICKING IN LATIN AMERICA
GLENN MORSE

Introduction identifies subject.

Introduction reveals objective tone to be used throughout.

The Reagan Administration has declared war on cocaine and has begun to take drastic steps to wipe out the drug. Some of these steps are aimed at Latin American countries that produce cocaine or are instrumental in transporting the finished product to the United States. Many of these governments welcome antidrug assistance from the United States, although most do so with differing degrees of reluctance.

Introduces general chronological strategy.

Use of APA documentation style.

Paragraph establishes the size of the problem.

As the cocaine epidemic mushroomed in the 1980s, the Reagan administration began an interdiction program to stop the flow of drugs into the United States, particularly Miami. More Drug Enforcement Agency agents were added, and ground crews were enlisted to inspect various parts and places of commercial aircraft, looking for hidden cocaine. Results were immediate. Twenty-five hundred pounds of cocaine were found in a shipment of Valentine's Day flowers on an Avianca Airline jet from Bogotá, Colombia (Anderson, 1985, p. 15). Eastern Airlines was rocked by the discovery of a drug-smuggling ring among its baggage handlers in Miami and Colombia. The stream of cocaine being smuggled into Florida was reduced to a comparative trickle by 1985. Unfortunately, stopping the flow into Florida did not stop or slow the tide of cocaine into the United States. Between 59 and 78 tons of cocaine entered the United States in 1983. That figure rose by one third in 1984 (Anderson, 1985, p. 15) and continued to rise through 1985.

Transition between paragraphs.

Description of eradication program.

With the results of the interdiction program mixed, the Reagan administration turned its focus to the source of the drugs. The administration introduced a coca eradication program in Latin America. Million-dollar grants were offered to induce coca farmers to destroy their coca crops and switch to legal crops. Money was also appropriated to help the governments of these nations to set up antinarcotic police forces to combat cocaine traffickers and to find and destroy cocaine-processing laboratories and related facilities. The governments of Colombia, Peru, and Bolivia, as well as other nations less tied to cocaine production, welcomed the aid. They were alarmed over the wealth and power of the traffickers and were being harassed by guerrilla movements and acts of terrorism that were financed by drug money (Anderson, 1985, p. 15).

The eradication program in Bolivia.

Bolivia is relatively new to the realm of cocaine-producing countries that are trying to eradicate coca production. Coca production accounts for nearly $600 million per year in Bolivia, three times the amount credited to Bolivia's official exports (Anderson, 1985, p. 17). Five percent of the 6.5 million people in Bolivia are involved in growing or processing coca. The farmers can earn as much as $3200 per acre per year by growing coca, while the government was offering $140 per acre in substitute crops (Robbins, 1986, p. 55). Former President Luis Garcia Meza and his chief of security were heavily involved in the cocaine trade in the early 1980s. But the government of new President Hernan Siles Zuazo has been working diligently to get the government out of the business (Anderson, p. 17). Serious attempts were made to try to meet both drug enforcement and eradication targets.

Use of specific figures builds credibility.

Results of the Bolivian program and elite force, the Leopards.

An elite Bolivian police force, the Leopards, was formed, with the use of American antidrug aid, in order to fight drug trafficking. This group has so far produced few positive results. Bolivia had already lost $7.2 million in U.S. aid in 1986 for failure to meet Hawkins bill requirements of 10,000 acres for crop eradication in 1985 (Robbins, 1986, p. 55). After warning of a further holdup of $58 million if their drug-fighting record did not improve, there was a flurry of army and police raids on suspected cocaine traffickers. The results were not impressive; not a single acre of coca plants was destroyed (Anderson, 1985, p. 18). To add further embarrassment to the Bolivian government, a unit of

245 Leopards was put under siege by 17,000 irate coca farmers, led by narcotics traffickers in the Chapare region. The farmers were protesting the rape of one of the women workers by two drunk Leopard members. After three days, with the Leopards nearly out of food and water, the Bolivian government was able to negotiate a peaceful settlement and rescue the unit (''Going to the Source,'' 1986). Other attempts at eradicating drug crops and production centers were met with stiff resistance by farmers and traffickers alike.

Facts and figures, and anecdotal information support this paragraph.

Operation Blast Furnace was initiated to help the Bolivians achieve more successful results in their drug-fighting campaign. Blast Furnace represented a ''significant escalation in an open-ended commitment'' to the use of the military against cocaine trafficking outside the United States (DeMott, Seaman, & Stanley, 1986, p. 12). American troops would stay in Bolivia for two months, transporting the Leopards to mission sites, where they would find and destroy cocaine-processing laboratories. The Bolivian government had received a promise from the United States that this operation was aimed at the destruction of the cocaine-processing centers and not at the coca farmers themselves. The coca plants, so vital to the Bolivian economy, would be left alone (DeMott, et al., p. 12).

Introduces second Bolivian program.

Controversy and bad luck shadowed Operation Blast Furnace from the beginning. In the first two weeks of the operation, only two laboratories were located and destroyed, and no large-scale traffickers were apprehended. Despite questionable results, the operation was described as a success (DeMott et al., 1986, p. 13) by the Bolivian and American governments.

End of Bolivian section of paper.

Ironic comment attributed to source.

In Peru the problems facing the government of President Alan Garcías are even greater than in Bolivia. The farmers in Peru are entrenched in the coca-raising business, and have stubbornly fought attempts to change Peruvian agriculture over to more acceptable crops. The farmers generally

Paper is progressing in order of importance (or size): Bolivia, Peru, Colombia.

support the Sendero Luminoso (Shining Path) Maoist group that preaches a communist revolution in Peru (Iyer, Beaty, Dietrich, & Scott, 1985, p. 31). They and a smaller guerrilla movement, the Puka Lactas (Red Fatherland) have continually harassed eradication teams as they worked to destroy coca fields.

Introduces political connection; revolutionary groups supported by drugs.

An elite police force, the U.S.-financed, 220-man Rural Mobile Patrol Unit (Iyer et al., 1985, p. 32), was organized to fight narcotics trafficking and aid eradication programs. The unit was sent to the interior of Peru to force coca farmers to join U.S.-backed crop eradication programs. Their alternative was to have the government confiscate their lands. This use of strong-arm tactics pushed the farmers even closer to the Shining Path guerrillas (Anderson, 1985), who are closely tied to narcotics traffickers (Taylor, 1985, p. 72).

Use of three different sources in this paragraph gives credibility through documentation.

Transition between paragraphs.

Despite these setbacks, President Garcías has escalated his war on cocaine. Police commandos based in Iquitos attacked 12 clandestine airstrips in the Amazon region. Operation Condor, a U.S.-backed paramilitary mission using only Peruvian troops, launched three assaults from August of 1985 until August of 1986 (Thomas, 1986, p. 72). These raids destroyed 270 labs, disabled 144 airstrips, seized more than 46 tons of cocaine paste, confiscated 14 planes, and led to the arrest of 1200 suspected drug traffickers. In 1985, 7000 hectares of coca-producing plants were eradicated (''A Condor Strikes,'' 1986, p. 23), 1000 more than their U.S.-imposed target (Thomas, p. 79) and double the amount in the previous five years. Still, 120,000–150,000 hectares were harvested, enough to supply the entire U.S. market (''A Condor Strikes,'' p. 23).

Though effective, the Peruvian program cannot keep up with drug traffic.

Colombia has established itself as the leading Latin American country in the fight against drug trafficking, probably because it is the country with the most problems with organized narcotics traffickers and insurgent groups backed

Introduces the biggest problem last.

by drug money. Seventy-five percent of the cocaine that reaches the United States is refined in Colombia (Anderson, 1985, p. 17). Colombia's drug processing and trafficking organizations are run by several moblike ''families'' whose leaders are known as narcotráficos (Iyer et al., 1985, p. 31). These ''drug kings'' have created the largest chemical exporting operation in the history of South America (Iyer, et al., p. 29). They are seen as modern-day Robin Hoods by the Colombian peasants who benefit from sharing the wealth of the narcotráficos (Anderson, p. 15).

Drug trafficking benefits Colombian peasants.

There are four major insurgency groups operating in Colombia. The oldest and best equipped is the Revolutionary Armed Forces of Colombia (FARC). It boasts 2000 active members and 3000 supporters. It is a rural movement based in the major coca- and marijuana-producing areas. The best-known of the antigovernment movements is probably the 19th of April Movement (M-19). It is an urban movement that has 900 active members organized into 140 cells. This is also the most dangerous group, linked to the Cuban government (Taylor, 1985, p. 71), and backed by Carlos Lehrer (Anderson, 1985, p. 15), narcotráfico número uno. Smaller groups are the National Liberation Army (ELN), with between 300 and 800 members and the Popular Liberation Army (EPL), with its dissident faction, the Pedro León Arboleda (PLA). Each of these groups claims around 250 members (Taylor, p. 71). All these groups receive protection money, either by extortion or voluntary donation, from the coca growers. Some act as informers to warn growers of impending raids by the government.

Political groups are linked to drugs. Paragraph reveals depth and complexity of the problem.

Many facts and figures give this paragraph power.

The government itself has been embarrassed by narcotics trafficking activities from within. By 1985, 100 air force personnel and two national policemen had been discharged from service for involvement in the narcotics trade. Four hundred judges have been investigated for complicity in the drug trade. Roman Medina, President Betancur's personal press secretary, was arrested for

Transition between paragraphs.

Student's style uses relatively short, very readable sentences.

smuggling 2.7 kilos of cocaine into Spain in diplomatic pouches (Whitaker, Shannon, & Moreau, 1985, p. 21). The fact that these dismissals, investigations, and arrests have been made is an indication that the Colombian government is seriously trying to remove corruption from its midst.

Paragraph reveals corruption related to drugs.

To counteract the destabilizing forces the Colombian government has moved to the forefront of antinarcotics activity. Since 1980, a crack paramilitary force, the Special Anti Narcotics Unit (SANU), has been used to hunt traffickers on a full-time basis. It is composed of 1700 top army and police recruits using U.S.-made UH-1 helicopters (Whitaker et al., 1985, p. 21). Since the assassination of Justice Minister Lara, 1500 raids resulting in 1425 arrests, have been carried out (Thomas, 1985, p. 50). In 1984, Colombian police seized 40,000 pounds of cocaine paste and finished powder. They destroyed 27.5 million coca plants and 262 laboratories (Anderson, 1985, p. 17). As of 1985, Lehrer and Escobar had been forced underground. Ochoa and Gilberto Rodriguez had been arrested in Spain on narcotics charges and were being held there. Botero was on trial in Miami for cocaine trafficking, one of the first to be extradited by the Betancur government (Whitaker, p. 21). Still, cocaine production was up 30 percent in Colombia.

Paragraph shows how stubborn the problem is. Despite efforts by Colombian government, production is up.

Despite, or perhaps because of, antinarcotic police work in countries such as Colombia and Bolivia, cocaine production has spread to other Latin American nations. Brazil and Argentina are becoming production and refining areas. Ecuador, once merely a transit country, has had an alarming rise in coca production (Thomas, 1986, p. 79). The Tranquilandia raid netted a record ten tons of cocaine, but within six months the cocaine supply was back to normal. Other countries took up the slack. Some of the refineries were moved to Bolivia and Peru, partly precipitating Operations Blast Furnace and Condor (DeMott et al., 1986, p. 13).

Paragraph shows the problem is spreading to other countries.

Congress, with then Speaker of the House Tip O'Neil leading the way, called for a ''multibillion-dollar outlay'' for a comprehensive antidrug bill (Duffy & Robbins, 1986, p. 6). Nothing has come of this call, nor is anything likely to because of budget cuts under the Graham-Rudman Act (''Going to the Source,'' 1986, p. 8). The key to effectively stopping the tide of drugs pouring into this country is not supply but demand. Until people, particularly the youth of America who are now experimenting with the cocaine derivative, crack, realize the consequences of their behavior, little can be done to control the problem. As Guillermo Angulo, Consul General for Colombia, observed, ''Colombia will continue to produce the crops as long as there are consumers. When there are no consumers, there will be no production'' (Anderson, 1985, p. 15).

Paper concludes interdiction and eradication programs are not the answer. The consumers are the real problem.

References

Anderson, H. (1986, 25 February). The evil empire. Newsweek, pp. 14–18.

A condor strikes the snowbirds of Peru. (1986, August 18). Newsweek, p. 23.

DeMott, J. S., Seaman, B., & Stanley, A. (1986, July 28). Striking at the source. Time, pp. 12–14.

Duffy, B., & Robbins, C. A. (1986, August 4). A jungle war at home and abroad. U.S. News & World Report, pp. 6–7.

Going to the source. (1986, February 3). New Republic, pp. 7–8.

Iyer, P., Beaty, J., Dietrich, B., & Scott, G. (1985, February 25). Fighting the cocaine wars. Time, pp. 26–35.

Robbins, C. A. (1986, July 28). U.S. mission: Cut off drugs at the source. U.S. News & World Report, p. 55.

Taylor, C. D. (1985, August). Links between international narcotic trafficking and terrorism (Department of State Bulletin) pp. 69–74. Washington, DC: U.S. Government Printing Office.

Thomas, J. R. (1985, June). International campaign against drug trafficking (Department of State Bulletin). Washington, DC: U.S. Government Printing Office.

Thomas, J. R. (1986, April). Narcotics control in Latin America (Department of State Bulletin). Washington, DC: U.S. Government Printing Office.

Whitaker, M., Shannon, E., & Moreau, R. (1985, February 25). Colombia's kings of coke. Newsweek, pp. 19–22.

APA-style references.

News magazines and department of state bulletins make credible sources.

The Process of Writing a Report

You must develop a thesis you can manage in a (relatively) short report. Note that Glenn does not attempt to write everything possible about drugs. He has narrowed his thesis to a subtopic of drugs: America's war on drug traffic from South America. (See Chapter One.)

How much information you need depends on your situation. If you want your reader to understand how big and how serious the problem is, as Glenn does, you must write a comprehensive report, incorporating as many facts as you can find. You will need to read many articles for this kind of report.

The simplest organizational plan for this kind of report is chronological: start at the beginning and then describe each step as it happened. Since the point of this kind of report is that it compiles data, you must cite information and give a reference for each fact you use. You must write a readable composition, not a list. Avoid too many obvious time signals like "first," "next," "then." Revise until each paragraph is clear and unified; edit until each sentence is readable and polished.

ACTIVITY 51 The United States Senate is investigating your research topic. As an expert on the topic, you have been asked to submit a report on its history (a history of women in the army, for example) similar to Glenn's report on cocaine trafficking. Your report should be thorough but concise.

WRITING AN ARGUMENT

Argumentative writing, in its broadest definition, includes any writing in which you attempt to convince your readers. Often argumentative writing includes "persuasive" writing. The difference lies in the difference between convincing and persuading. Some people believe they can be persuaded to do something even though not convinced that it is the right action. Persuasion, they say, relies on appeals to emotion and character, whereas argumentation should rely on reason alone. The truth is that often persuasion and argumentation occur together.

However, in most research writing, tradition calls for a heavy reliance on reason and factual data. A scientific argument presents a thesis—some conclusion the researcher holds (that cigarettes cause cancer, for example)—and then presents reasons and examples that support the thesis.

Read the following example of an argumentative paper. What kinds of appeals does Cindy use? What kind of data does she have? Is there enough data to make her point? Does she convince you? Note the use of traditional endnote numbers and references.

1

THE TORTURING OF LAB ANIMALS
CINDY GILPIN

Laboratory animals are used to test the effects of almost all new chemicals. They are used by industry to test cosmetics, by medical researchers to predict how humans may react to certain drugs and medicines, and they are also used in high school biology for experiments and science fair projects. Advocates of animal rights are trying to stop the use of laboratory animals because many of these experiments are unnecessary and painful to the animals. Researchers, however, say that their experiments are vital for human safety.

Introduction introduces argument.

Mice, rats, rabbits, and monkeys are bred at special farms specifically for experimental use because their biology is similar to ours. Testing drugs or cosmetic products on laboratory animals first, it is believed, may avert harm to people. Researchers say that the experiments are carefully conducted so that animals are not harmed. The animals are used in medical schools in order to allow students to get used to performing operations on living organisms before beginning on a human. Experiments in medical schools and biology classes are constantly repeated to allow all students to see the results firsthand, because it is believed this is the best way to learn.

Paragraph presents strategy: begin with weaker side.

Objective presentation of arguments.

Researchers claim that small amounts of the compound that causes cancer in laboratory animals might cause the cancer to appear in humans. So many experiments, the researchers claim, help to protect mankind. Animal rights advocates also complain about the excessive number of animals used in a single experiment. Researchers insist that by using a large number of animals per experiment, they are more certain of their results and can predict the percentage of people who will be affected.

Student makes concessions to the weaker side.

Advocates of animal rights feel that laboratory animals should not be used in experimentation, that a substitute should be

Presentation of stronger side.

found. These advocates claim that laboratory animals are misused and mistreated all the time, from high school biology classes to research laboratories. High school classes urge students to compete in science fairs. The projects entered in these fairs torture the experimental animals. One high school student ''cut off the legs and tail of an amphibian to show that the tail will grow back, but the legs won't. This is a known fact that can be found in most biology books.''[1]

Endnote documentation style.

Medical schools are also under attack. ''Dogs are fed strychnine and rats are injected with diseases, and the students are supposed to observe their responses.''[2] It is good for medical students to be able to observe these results firsthand, but to keep repeating the experiments year after year is unnecessary. Advocates argue that students would learn just as much if shown a series of films. But repeated experiments are defended on the claim that ''seeing a case and reading about it [or passively viewing it] are two different things.''[3]

Student maintains objective stance: presents both sides.

Animal righters are also after the cosmetic industry. Every time a new cosmetic is invented, or a new ingredient is added to an already existing product, it must be tested. Thousands of animals are needlessly tortured every year because of this. These animals are frequently used to test eye irritation. ''A few drops of the product are dripped into the sensitive part of the corneas of rabbits.''[4] The tests are continued until nothing happens or the rabbit goes blind. ''The tests cause extreme suffering, and the companies do not give pain relieving drugs.''[5] The researchers say pain killers will affect the outcome of the experiment with false information.

Use of quotation to illustrate, substantiate points.

The efficiency and accuracy of the tests and experiments are also reasons for not using lab animals. In toxicological studies, huge doses of a substance are fed to an animal, and the results are measured for the possible threat to humans. For example, about two years ago, a Canadian study indicated that saccharin caused bladder cancer, and the FDA was planning to ban its use. In the tests, ''rats were fed the daily equivalent of the saccharin in 800 cans of diet soda.''[6] Because the amount of saccharin consumed per day is so large, acontroversy arose. The FDA didn't ban the use of saccharin, but posted warnings on products containing saccharin. When excessive dosages of a product are given to an animal, the credibility of the results is questionable. Other experiments involve alcohol and cigarettes. Mice are fed only alcohol and are kept constantly drunk. The effects of this alcohol upon the biosystems of the mouse are then examined to determine the effect of alcohol upon a human. Mice are also used in cigarette-smoking experiments. Mice are hooked up to an apparatus for a certain amount of time every day. This apparatus forces them to inhale cigarette smoke. The effect of this upon mice is then studied and the results are projected for humans.[7]

Facts and figures, use of documentation give credibility to the argument.

Animal rights advocates feel that much of the repetitious testing could be eliminated if the test animals were treated differently. The amount of fear in an animal can and does affect the outcome of experiments, which must then be repeated often to try and find an accurate result. The researchers treat the animals badly. Mice and rats are picked up by their tails. Instead of being placed on the lab table, the animals are dropped onto it. ''The animals are kept isolated,''[8] and because of this they develop sicknesses or other side effects that are mistakenly attributed to the product being tested.

Paragraph presents a counter-argument.

Cumulative impact of several paragraphs makes this the stronger side.

No solution to the problem of whether animals should be used in lab experiments will satisfy everyone. However, some changes should be made. Unnecessary experiments for science fairs should be eliminated. Medical students may need to experiment on animals, but some of this knowledge can be acquired through film and observing others. Researchers should treat the animals better by keeping them in better living conditions and by not dropping them. With cooperation on both sides, laboratory animals could still be used for experimentation without being tortured.

Conclusion takes reasonable position, calls for cooperation of both sides.

Endnotes

Endnote style.

1. E. M. Leeper, ''The New Methods May Reduce Needs, Can't Replace Laboratory Animals,'' BioScience, Jan. 1976: 9–12.
2. Peter Gwynne with Sharon Begley, ''Animals in the Lab,'' Newsweek, 27 Mar. 1980: 84–85.

Shortened references.

3. Gwynne, 85.
4. Gwynne, 85.
5. Gwynne, 84.
6. Gwynne, 85.
7. ''Mighty Mice,'' Time, 27 Mar. 1980: 51.
8. Robin Maranty Henig, ''Animal Experimentation: The Battle Lines Soften,'' BioScience, Mar. 1979: 145–48, 195–96.

Bibliography

Gwynne, Peter, with Sharon Begley, ''Animals in the Lab,'' Newsweek, 27 Mar. 1980: 84–85.

Henig, Robin Maranty. ''Animal Experimentation: The Battle Lines Soften,'' BioScience, Mar. 1979: 145–48, 195–96.

Leeper, E. M. ''The New Methods May Reduce Needs, Can't Replace Laboratory Animals,'' BioScience, Jan. 1976: 9–12.

The Process of Writing an Argument

Research arguments are similar to critiques and comparisons; the arguments derive from published sources. That is, the arguments tend to be factual rather than emotional. You must argue, as Cindy does, that one source says X, but another source says *not* X. You can be entirely impartial, drawing no conclusion—in effect, you permit your readers to draw their own conclusions—or you may organize your paper to emphasize the side of the argument the data shows is correct.

You should start by telling the reader what the argument is, by giving some background to the argument. If there is a weaker or "wrong" side of the argument, you should present that side first. It is well to begin by making concessions; present the weaker side in its best light. Then present the evidence for the other side. The more evidence and documentation you can present on both sides, the more convincing you will be. (See Chapter Six).

The tone of a scientific or academic argument is important to your credibility. Avoid sarcasm; avoid all signs of unfairness or bias. You should write in the tone of an impartial judge examining the evidence in a dispute. Your job is not to win but to teach your reader how to understand the argument.

ACTIVITY 52 Write an argumentative paper based on your research topic. Limit your thesis and then present arguments and data illustrating the thesis. You are assistant to an attorney who is presenting a major case before the United States Supreme Court. Your job is to put together a thorough and carefully documented argument which the attorney can use as a source in preparing his/her case.

WRITING THE RESEARCH PAPER

Research papers include most of the writing situations already described in this section. They can include the report of an experiment, a review of literature, or just any long report. What is characteristic of most research papers is that they tend to be relatively longer than other papers. There is no rule about this, and some research papers can be only a few pages long, but it is not uncommon for research writing to be a dozen or two dozen or several dozen pages long. Especially experimental research reports tend to be long because researchers try to give as much information as possible about the experiment. The other criterion is that most research papers require documentation.

The following example is a review of literature investigating the question of whether apes can talk. Note the use of MLA style documentation.

Optional title page

Can Apes Talk? A Review of Selected
Literature in the Ape-Language
Research

by

Lisa McLaine

Advanced Composition 201
Professor Horton
April 25, 1987

CAN APES TALK?

Required formal outline

I. Background
   A. Significance
   B. Ape communication in the wild
   C. Early efforts to teach apes to talk
      1. Viki and speech
      2. Ameslan
   D. Research problem: definition of ''language''
      1. Concrete versus abstract language
      2. Ability to refer: naming versus requesting
      3. Ability to make a sentence: stringing words together

II. A History of Research
   A. Washoe: first signing ape taught ASL
      1. Swearing
      2. Negatives
      3. New uses for words
      4. Inventing words
      5. Washoe's long sentence
      6. Washoe's significance
         a. First experiment with scientific controls
         b. Washoe's use of language similar to that of children
   B. Lucy (ASL) Fouts
      1. Motivation for language use
      2. Understanding spoken English
      3. Humor: Lucy and Washoe
      4. Inventing words
      5. Expressing emotion
   C. Sarah (plastic tokens), Premack
      1. Reading and writing
      2. Same/different tests
      3. Sarah's compound sentence
   D. Lana (Yerkes: computer), Rumbaugh
      1. Reading and writing
      2. Lana's sentences
         a. Grammatically correct
         b. Imitated by computer program
      3. Lana's success

E. Sherman & Austin (Modified Yerkes), Savage-Rumbaugh
   1. Sharing work and reward versus conditioned behavior
   2. Sorting test
   3. Inventing new use for a word
   4. Using language to ''refer''
      a. Long, arduous training
      b. More complex than pigeon behavior
F. Koko
   1. The only gorilla in ape-language research
   2. Koko's linguistic ability
      a. Spontaneous use of 150 signs
      b. Doubt about meanings
      c. Koko's MLU
      d. Humor: Koko's joke
G. Ape researchers optimistic

III. Critical views
A. Nim (ASL) Terrace
   1. Replicating Gardners' study
   2. Animals cannot use language
      a. Apes cannot make sentences
      b. Words have no meaning
      c. Ape behavior is rote, like pigeons'
   3. Nim imitates, interrupts teachers
   4. Alll apes imitate teachers
B. Imprecise Reporting
   1. Naming versus selecting
   2. Mistakes rationalized
   3. Faulty recording procedures
   4. Ambiguous use of ''spontaneous''
C. Design Problems
   1. Treatment of mistakes
      a. Washoe's
      b. Lana's
      c. Other researchers

- D. Interpretation problems
  - 1. Occam's razor
  - 2. Too much anecdotal information
    - a. Informal data collection methods
    - b. No control for subjectivity
      - (1) Koko's imitations
      - (2) Koko's pig Latin and rhymes
      - (3) Doubts about Koko
    - c. No tests for meaning
  - 3. Trainers edit responses
  - 4. Serious doubts about entire line of research
- E. Cuing: The major flaw
  - 1. Clever Hans error
  - 2. Apes dependent on cues
  - 3. Some researchers object, deny cuing

IV. Results
- A. Review of key points
- B. Severe criticism: effect on research
- C. Apes cannot use language
- D. Researchers bitter, tired of this question

V. Works Cited

Introduction presents thesis statement or question.

Introductory strategy: historical background.

Significance of this research.

MLA-style documentation.

Two sources validate the point.

The research contains a problem, which the student treats objectively. Student clarifies the thesis question.

Can Apes Talk? A Review of Selected Literature in the Ape-Language Research

Background

Can apes talk? We know apes communicate with each other in the wild (Premack, A. 5-8). Could they be taught to speak? In 1951, the chimpanzee Viki was taught ''mama,'' ''papa,'' ''cup,'' and ''up,'' but Viki never used these ''sounds'' like words (Reynolds 219). The ape, it seems, is incapable of speech, and Viki's was the last speech experiment (Ristau and Robbins).

The ape-language research was undertaken with high hopes. It was thought this research might help to ''better define the fundamental nature of language'' (Premack and Premack 92). Further, the research may shed some light on ape and/or human intelligence. Some researchers hope the nonspoken language systems may work with retarded or brain-injured people (Ristau and Robbins). Terrace believes (Nim 4) this work may have implications for meeting extraterrestrials: learning how to communicate with alien minds. This research may help us understand mankind's relation to the other animals. Is the human animal a ''special creation'' with a unique capacity for language? The possibility of language in a nonhuman makes Homo sapiens seem less unique. (Premack, D. Gavagai!; Terrace, Nim.).

A key problem in this research is the definition of language. Many animals can learn words for physical objects or behaviors, like ''sit'' and ''fetch.'' But can animals learn words like ''Europe'' or ''democracy''? Humans use language to ''refer''—to name things without requesting them (Terrace, ''In the Beginning Was the 'Name'''). Can apes do this? Can an ape make a sentence (Terrace, ''Why Koko Can't Talk'' 8)? Many language tests have been suggested, and most of them have been passed by one or more apes (Premack, D. Gavagai 124; Ristau and Robbins 148). Given some method of communication other than speech, can apes use language?

A History of Research

Washoe In 1966 Beatrice and Allen Gardner began to teach American Sign Language to Washoe. In three and a half years, Washoe learned 132 words (Terrace, Nim 10).

Washoe has several language abilities. For example, Washoe is reported to swear. She had been taught the word ''dirty'' in reference to the toilet, but she began to apply it to a trainer. She signed, ''Dirty Jack gimme drink.'' (Linden, Apes, Men, Language 8). Washoe once expressed a negative, and she does other things with language she was not taught to do. She was taught to make the sign for dog when she sees a dog, but Washoe also signs ''dog'' when she only hears one (Droscher 212). And Washoe ''invents'' new words. When shown a swan, Washoe replied ''water bird,'' a phrase she had not been taught. The Gardners report that Washoe uttered at least one longer sentence that was not a request; in a filmed sequence, Washoe says, ''Baby in my drink'' when she sees a tiny doll placed in her cup.

The Washoe experiment was the first study with scientific controls, and the ''extraordinary claims'' of the Gardners gave the impression that there was little difference between Washoe's language and that of a child using sign language. ''Indeed, in one study they found the ape superior to the child'' (Premack, D. Gavagai! 32).

Lucy Lucy was trained to use ASL by Roger Fouts. Lucy, like Washoe, lived with her trainers; the idea was to create a rich environment that would motivate the chimpanzee to use language (Premack, A.). In this environment, Lucy appeared to understand spoken English. Some researchers have reported that apes understand humor and laugh. In one example, Fouts pretends to swallow his sunglasses. Standing sideways to Lucy, Fouts opens his mouth and passes the sunglasses down the side of his face away from Lucy. ''Lucy thinks this is hysterically funny'' (Linden, Apes, Men, Language 97). Washoe has a raunchier sense of humor. ''One day while riding on Roger's shoulders, Washoe

Subheadings optional in MLA (required in APA).

Information about Washoe is given without counterargument. Student is making the case for the weaker side.

Washoe material is well documented.

Anecdotal evidence throughout this paper helps maintain reader interest.

pissed on him and then signed 'funny' in a self-congratulatory way'' (Linden, Apes, Language, and Men 97). Lucy, like Washoe, invents words (Linden, Apes, Men, Language 109), and Lucy is said to express emotions. She signed ''cry me, me cry'' as she watched her foster mother leave (Linden, Apes, Language, and Men 111).

Lucy evidence is presented without counterargument.

Sarah David Premack taught the chimpanzee Sarah to ''read'' and ''write'' a language consisting of plastic tokens in random shapes. Each token represented a word: the token for ''apple'' was a blue triangle (Gill and Rumbaugh). Some of Premack's claims for the apes' abilities are amazing. He says they can match $3/4$ of an apple with $3/4$ of a cylinder of water (Premack, D. ''Possible General Effects'' 268). These tests usually involved the apes selecting either ''same'' or ''different'' when shown two objects. Sarah's most impressive performance is her comprehension of a ''compound'' sentence: Sarah insert apple pail banana dish. Sarah had to understand the relationships in this sentence in order to get the food in the right containers (Premack and Premack 99). Other researchers are more critical of Sarah.

Student expresses amazement at claims by Premack, but does not contradict the author nor offer counterevidence.

Lana Lana was taught a special language, ''Yerkish,'' by David Rumbaugh at the Yerkes Center.[1] The language was made up of geometric patterns (lexigrams) on a computer keyboard. Lana could write sentences like ''Please machine give banana,'' (Rumbaugh and Gill 165). Her lexicon (word list) had nouns and verbs, mostly requests, and a few attributes, altogether 150 words.

Lana's sentences are grammatically correct (Von Glasersfeld 127), but because her sentences are all very short and similar, it was possible to create a computer program that could imitate Lana's performance. Nevertheless, the Lana researchers overall were enthusiastic about apes' language abilities (Rumbaugh and Gill; Savage-Rumbaugh and Rumbaugh 249).

The style of this paper is very readable. The language is simple, even though the sentences are relatively long.

Sherman and Austin Sue Savage-Rumbaugh designed an experiment with two chimpanzees,

Sherman and Austin (Ape Language). Like Lana, Sherman and Austin used computers with Yerkish symbols. Much of the research with apes has been criticized as simple stimulus-and-response behavior; if apes get rewards for pushing computer keys, they will learn to push them, as pigeons can learn to do just about anything for rewards. But Savage-Rumbaugh et al. set up an experiment in which the two apes had to work together. One chimp has access to a food site but must request a tool from the other chimp. The chimps share the reward (''Can Chimpanzee Make Statement?'' 479).

After Washoe, Lucy, Sarah, and Lana, a more sophisticated experiment is introduced.

In another test, the chimps learn to sort words (computer lexigrams) into groups—food words, tool words. Some researchers doubt the apes' sorting ability: there are not very many items to sort, mostly edibles; apes may simply discriminate edible versus everything else (Ristau and Robbins 207).

Savage-Rumbaugh reports one occasion in which Austin invents a new meaning for a word (Ape Language). Austin uses ''scare'' when he sees attendants in white coats carrying an anesthetized chimp past the window. ''Scare'' had been taught as a request to play a game in which the teacher pretends to scare chimps. Did Austin invent a new use for ''scare''? Sherman and Austin did, eventually, learn to ''name'' or ''refer'' to objects, but only with ''long and arduous'' training (Savage-Rumbaugh et al., ''Can Chimpanzee Make Statement?'' 485). Savage-Rumbaugh says Sherman and Austin's behavior was highly complex, different from the simple peck-and-reward behavior of pigeons (Savage-Rumbaugh, ''Verbal Behavior'' 248). The apes had to use a tool to get a reward; one knew where the reward was, one had access to the tool. She concludes that apes can learn sophisticated language behaviors (Savage-Rumbaugh, Ape Language 379).

References to three different sources by the same researcher show the student is familiar with the relevant research.

Koko Koko is the only gorilla in this research. Koko was exposed to both ASL and oral English. Koko is reported to have a large

The Koko material was saved for last because it is problematic and conflicts with other research.

vocabulary and more linguistic ability than other apes (Patterson 72). To date Patterson has published many anecdotes about Koko. (The Education of Koko and Conversations with a Gorilla) but has released very little research data.

Patterson reports that Koko ''frequently used spontaneously over 150 different signs'' (Patterson 77), and her trainers were in significant agreement about her signs. Other researchers say that it is not always possible to know what Koko's signs mean (Ristau and Robbins 173). Ristau and Robbins say that only the Koko research disputes findings by Terrace and others critical of ape language. For example, though Nim's utterances did not get longer after a certain age, Koko's did continue to increase in mean length of utterance (Patterson 92). The implication is that Patterson is the lone holdout, despite criticism, still claiming her ape uses language (Petitto and Seidenberg).

Like others, Patterson refers to humor and ''jokes'' by the ape. Koko signs ''Thirsty drink nose.'' Koko knows very well that ''drink'' has nothing to do with ''nose''—it is not a ''mistaken'' sign. And we are told Koko laughed. Patterson concludes that based on Koko's abilities, ''language is no longer the exclusive domain of man'' (95).

This side of the question concludes with affirmation: apes can use some forms of language.

Multiple reference.

Despite various critics and some obvious flaws in the earlier studies, the overall tone of these reports is positive. Each new study attempts to control for problems in previous studies. Researchers seem to be making progress. If we ask those who have worked with the apes, we will get a positive answer from most of them; so far this research suggests that apes can use some forms of language (Premack, A. 106; Ristau and Robbins 231; Savage-Rumbaugh and Rumbaugh, 305).

Critical Views

Nim Herbert S. Terrace, after criticizing the Washoe study, attempted to construct a more carefully controlled experiment, using the chimpanzee Nim.[2] Like Washoe, Nim was taught ASL, but unlike Washoe, he was kept in special quarters where he could be closely monitored. At first, Terrace was optimistic in his research. Nim learned a vocabulary of over 100 words and had begun to string them together in sentences. However, after an analysis of 19,000 videotaped utterances, Terrace concludes that apes cannot make sentences (Terrace, ''How Nim Changed My Mind'' 66). According to Terrace, all the ape-language research is seriously flawed; he does not believe apes are capable of language (Terrace et al., ''Can an Ape Create a Sentence?''). Terrace believes ''words'' have no meaning for apes: they are just nonsense symbols the ape learns to manipulate to get its reward. The video record shows Nim tries to grab things, only resorts to sign language when nothing else works, and then he imitates his teacher's signs. ''Even though Nim was stringing together combinations of signs, he knew no grammar'' (Terrace, ''Why Koko Can't Talk'' 9).

After presenting the weaker side of the argument objectively, the student presents the stronger side.

Raised number indicates a content note at end of paper.

The student uses a combination of paraphrase and quotation.

Terrace cites research in which pigeons cooperate for rewards; one tells the other which lights are on. This kind of behavior, he says, is simply rote learning, and so is the behavior of the chimpanzees (Terrace, ''In the Beginning''). A significant fact in his own research, he says, is this:

> None of the features of Nim's discourse—his lack of spontaneity, his partial imitation of his teacher's signing, his tendency to interrupt—had been noticed by any of his teachers or by the many expert observers who had watched Nim sign (Terrace, ''How Nim Changed My Mind'' 75).

Indent 10 spaces for a quotation longer than 4 lines.

Attribution within sentence constitutes documentation.

Terrace et al. conclude that all the other apes were prompted by their trainers. The trainers induce the apes to sign by signing first.

Imprecise Reporting There is criticism about the way the ape-language research has been

Ape research is flawed with inexact language.

reported. Much of the information has been reported ''in a rather cursory manner'' (Savage-Rumbaugh et al., ''Do Apes Use Language'' 49). For example, researchers report that animals ''named'' food, ''asked for'' activities—but no such ''naming'' or ''asking'' took place. ''When Sarah selected [my emphasis] the chip that Premack called 'candy,' she was said to have named and asked for the candy'' (Savage-Rumbaugh et al., ''Do Apes Use Language'' 50). And frequently when chimps select things, they are given only two choices, occasionally no choice; the experimenter holds up an apple and the ape ''selects'' the only chip on the table, the ''apple'' chip. Terrace says in too many tests Sarah selects (she does not ''supply'') one answer from a pair of alternatives (Terrace, ''Problem Solving'' 170).

Quotation to illustrate researchers' rationalization.

Sarah's ''mistakes'' are rationalized. Premack says, when ''bored'' Sarah would steal all the tokens and form sentences on the floor, answering all the questions asked of her (Terrace, ''Problem Solving'' 174). He feels this indicates the ape's willingness to use language. But others feel ape mistakes are being disregarded (Terrace, ''Problem Solving'' 174). When they tried to get Sarah to choose between two tokens (one of which named the fruit reward) before being allowed to eat the reward, ''Surprisingly often . . . she chose the wrong word. It then dawned on us that her poor performance might be due not to errors but to her trying to express her preferences in fruit'' (Premack and Premack 97).

Criticism of inexact data.

Trainers kept no records of how many times the chimp tried, what percentage of tries were correct, and so on. Terrace says both more drink and drink more were recorded as correct responses in the Washoe tests (Terrace, ''How Nim Changed My Mind'' 68). ''Spontaneous'' came to mean answering questions. And if Washoe signed cat when shown a picture of a cat, her sign was recorded as ''spontaneous.'' In some research ''spontaneous'' meant only that the ape was not ''prompted'' with a stimulus to produce the sign the trainers expected (Ristau and Robbins 157). Terrace's analysis of the Washoe film shows that under prompting from her trainer (''what that''), Washoe first signs ''baby,'' then with more prompting she says ''in,'' and with still more prompting she says ''my drink.'' Terrace says this is neither spontaneous nor creative use of language (''How Nim Changed My Mind'' 76).

Design Problems The Gardners used strange rules by which any color is counted correct when Washoe is asked what color the ball is. As long as Washoe got the right category of, for example, grooming aids, she was counted correct whether she signed ''comb'' or ''brush.'' And on the basis of such evidence they concluded that Washoe's performance is comparable to that of a human child (Terrace et al. 900). Washoe's mistakes are discounted by the Gardners.

Researchers criticize the design of the Washoe experiment.

Lana's errors, too, make a problem. Lana is said to ''erase'' sentences that displease her, with the result that negative data are simply eliminated (Ristau and Robbins 183–84). ''This is an extremely serious criticism of the Lana project, which makes it almost impossible to determine Lana's grammatical abilities'' (Ristau and Robbins 184). When Lana's answer to a question is wrong, the investigators ignore her answer and continue to question her. Her wrong answers are discounted entirely or are attributed to ''obstinacy'' or ''lying'' (Ristau and Robbins 184). ''All ape researchers discard repetitious utterances; so while we are told they were numerous, we cannot know how numerous nor can we analyze them in any way'' (Seidenberg and Petitto, ''Ape Signing'' 123).

A serious criticism is raised concerning negative data.

Researchers all tend to ignore Occam's razor —they accept the ''richest'' interpretation without any attempt to evaluate simpler interpretations (Seidenberg and Petitto, ''Ape Signing'' 124). Washoe's swan is an example; it is possible that Washoe only meant that she saw water

Researchers ignore Occam's razor.

The Koko research illustrates excessive use of anecdotal data.

and she saw a bird (Premack, A. 74). We need more data about her use of signs. Overall, there is too much anecdotal information, not enough hard data. Patterson is particularly guilty. There is no control for subjectivity. Trainers who are intensely involved with animals do not make good observers (Linden, Silent Partners 129). Terrace says the film Koko, A Talking Gorilla shows the ape imitating her trainer (Terrace, ''Why Koko Can't Talk'' 9), and other claims by Patterson are ''wholly implausible.'' For example, Patterson tells us Koko can understand pig Latin; she correctly identifies ''andy cay'' as candy (9) and Koko can rhyme. But Koko cannot make human sounds and does not know the alphabet either orally or written, so such a claim seems very unlikely (Terrace, ''Why Koko Can't Talk'' 9). There is a suspicion that ''most of the accomplishments claimed for Koko are not authentic'' (Linden, Silent Partners 121).

Researchers question the logic of some claims about Koko.

Premack says apes sign very fast, are repetitive and redundant, and out of this garble, trainers edit an appropriate response (Premack, D. 32). Because apes are repetitious and redundant, anecdotes about word combinations are meaningless. Patterson for example has no data on whether Koko attaches different meanings to You listen me and Me listen you, and there is ''good reason to doubt'' what Koko's signs mean (Petitto and Seidenberg 180). Such data as Patterson does report ''are subject to alternative interpretations that do not require the radical conclusion that apes are able to learn aspects of human languages, particularly ASL'' (Petitto and Seidenberg 180-81).

Researchers alter data.

There are not enough data, not enough controlled testing in the ape-language experiments (Savage-Rumbaugh, et al. ''Do Apes Use Language'' 53). Even the Lana researchers have begun to doubt that their research proved very much, especially whether the apes' signs have any meaning, whether they are anything more than stimulus-response imitations (Ristau and Robbins 183).

Several researchers cast doubt on all ape-language research.

Cuing: The Major Flaw The major flaw in all this research is the ''Clever Hans'' [named after an ''amazing'' German Horse] cuing that was available to the chimpanzees (Chevalier-Skolnikoff 64-65). The chimpanzee ''hesitantly or sloppily or repetitiously or hurriedly executes some hand movement while closely watching the experimenter's face.'' The trainers, trying very hard to get the animal to perform, can hardly conceal their emotions as the animal makes the right or wrong choice. The chimpanzee becomes dependent on cues from the trainers, and as the trainers attempt to withhold obvious cues like facial expressions, ''the chimpanzee looks for more minimal cues'' (Savage-Rumbaugh, et al. ''Do Apes Use Language'' 54). Other researchers now suspect that the apes were being cued (Linden, Silent Partners 20).

The Clever Hans phenomenon. The major flaw is withheld until the end.

However, not everyone is as certain as Terrace is that Clever Hans explains much ape language. Sarah's tests used a naive trainer (who did not know the symbol language) (Premack, A. 102-03). Ristau and Robbins say that better tests now use double-blind experiments, in which the chimps cannot see the trainers. Ristau and Robbins conclude that the Clever Hans explanation is unwarranted (239). Savage-Rumbaugh says she found training her chimps so hard that it is not reasonable to assume they could easily learn through less obvious cues (Ape Language 77). She says ''cuing intentional or otherwise is not an effective way to teach chimps.'' But, in a re-analysis of the Lana project, Savage-Rumbaugh et al. admit ''we could not rule out the possibility of unintentional cuing'' (''Do Apes Use Language'' 54-55). And they conclude there is no ''definitive demonstration'' that Washoe, Sarah, Lana, Koko, or Nim used symbols representationally (55). The language-trained chimps were ''no more adept than wild apes, who use gestures for a variety of purposes (Savage-Rumbaugh, et al. ''Do Apes Use Language'' 60).

The issue of cuing remains controversial in the research.

Results

Conclusion begins with a summary review of the main points affirming ape language use.

The ape-language research began with optimism and a high level of public interest. The possibility that apes might be able to communicate with language appealed to the public, who were used to seeing performing—even talking—animals on television and in movies. Serious researchers quickly discovered that oral languages were impossible but that apes had a natural capacity for gestures. Early experiments showed that apes could learn sign language. Washoe used ASL to make requests, to swear, to invent names. Lucy, too, used ASL, and she also seemed to understand spoken English. The ape Sarah learned a new kind of language using plastic symbols, or tokens. Sarah's plastic sentences were easier to monitor than the gestures of the ASL chimps. Claims for Sarah were that she could not only read and write this plastic-token language, but that she understood concepts like ''same'' and ''different'' and could read compound sentences. Chimpanzees Lana and Sherman and Austin were taught to use computers with geometric symbols for words. The computer-using apes, too, proved very adept at language. Sherman and Austin learned to use their computer language to talk to each other and work cooperatively toward a goal they could only reach through the use of language. Koko, the only gorilla in the research, is said to be even more talented with language than the chimpanzees. Overall, each of the experiments with these animals reports good results. Each experiment attempts to make up for deficiencies in earlier experiments, and on the face of it the methods became more specific and more reliable as the research progressed. The general answer from these researchers is that indeed, apes seem to be able to use certain aspects of language.

After reviewing the affirmative side, the conclusion reviews the negative side.

However, this research was severely criticized when Terrace, working with the chimpanzee Nim, discovered that Nim was imitating his trainers. At first, like other researchers working with animals, Terrace had been optimistic about Nim's ability, but careful analysis of videotapes showed that Nim's signs were preceded by the trainer's signs. Analysis of films of other apes like Washoe, Koko, and so on showed the same thing. Like trained pigeons, the apes learned to perform for rewards.

Many research problems are cited.

Other criticisms of this research focus on the imprecise reporting of experiments, rationalizing mistakes by the apes, and failure to keep adequate records. Furthermore, some of the research was criticized for faulty research design, which failed to distinguish between ambiguous behaviors, discounted errors, and discarded repetitions. Many of the researchers are accused of overinterpreting the apes' behavior, failing to apply Occam's razor (that is, they accept the difficult conclusion that apes are using language and ignore a simpler conclusion—that apes are merely performing for rewards). In a few cases critics are highly skeptical. Koko, for example, could not possibly understand pig Latin or rhymes. And researchers edit apes' responses. Overall, critics charge there is too much storytelling, too many amusing anecdotes about apes and their trainers, and not enough hard data. The major flaw in this research, critics say, is the Clever Hans phenomenon: researchers attribute intelligence and linguistic ability to an animal, missing the fact that they themselves are cuing the animal. Though the Clever Hans effect has been disputed by researchers and later experiments attempt to control for accidental cuing, there is no denying that early experiments did not control for cuing—it is clearly visible in videotapes and film. The answer from critics of the research is clear: no, apes cannot use language.

Critics agree—apes cannot use language.

The effect of criticism on research.

The effect of criticism has been devastating on the research. Many researchers have been made to look ridiculous. Serious researchers—some with large reputations—have been shown to make the most elemental sorts of research errors. A few have been nearly accused of being fools or worse, frauds, doctoring the data.

Student agrees with negative answer to research question.

Sadly, it is difficult to conclude that this research so far reveals anything except that animals will perform very convincingly for rewards. Researchers who devoted years of their lives and staked their professional reputations on this research have come to the conclusion that it was all hopeless. Language belongs to the human race only. In 1986 Eugene Linden (Silent Partners) declared that enthusiasm for this research no longer exists. ''With few exceptions, the participants have been exhausted if not embittered by their experiences with sign-language experiments and the passions those experiments aroused. Few of them want to look back'' (10-11). So far the answer to this question seems to be negative. Apes can't talk.

Final quote saved for end.

Notes begin on a new page.

Notes

[1] Robert Yerkes established the first primate center in America: Yerkes Laboratory of Primate Biology at Orange Park, Florida.

Content notes, MLA style.

[2] Nim Chimpsky's name is a pun on the name of the linguist, Noam Chomsky.

Works Cited

Chevalier-Skolnikoff, Suzanne. ''The Clever Hans Phenomenon, Cuing, and Ape Signing: A Piagetian Analysis of Methods of Instructing Animals.'' The Clever Hans Phenomenon: Communication with Horses, Whales, Apes, and People. Annals of the New York Academy of Sciences Volume 364. Ed. Thomas A. Sebeok, and Robert Rosenthal. New York: New York Academy of Sciences, 1981. 60–93.

Gill, Timothy V. and Duane M. Rumbaugh. ''Training Strategy and Tactics.'' Ed. Duane M. Rumbaugh. Language Learning by a Chimpanzee: The LANA Project. New York: Academic Press, 1977. 157–62.

Linden, Eugene. Apes, Men, and Language. New York: E. P. Dutton, 1974.

——. Silent Partners: The Legacy of the Ape Language Experiments. New York: Times Books, 1986.

Patterson, F. ''The Gestures of a Gorilla: Language Acquisition in Another Pongid.'' Brain and Language 5 (1978): 72–97.

Petitto, Laura A., and Mark S. Seidenberg. ''On the Evidence for Linguistic Abilities in Signing Apes.'' Brain and Language 8 (1979): 162–83.

Premack, Ann J. Why Chimps Can Read, New York: Harper, 1976.

Premack, Ann James, and David Premack. ''Teaching Language to an Ape.'' Scientific American 227.4 (1972): 92–99.

Premack, David. Gavagai! or The Future History of the Animal Language Controversy. Cambridge: MIT P, 1986.

——. ''Possible General Effects of Language Training on the Chimpanzee.'' Human Development 27 (1984): 268–81.

Reynolds, Vernon. The Apes: The Gorilla, Chimpanzee, Orangutan, and Gibbon—Their History and Their World. New York: Harper, 1967.

Ristau, C. A., and D. Robbins. ''Language in the Great Apes: A Critical Review.'' Advances in the Study of Behavior Volume 12. Ed. Jay S. Rosenblatt, Robert A. Hinde, Colin Beer, Marie-Clarie Busnell. New York: Academic Press, 1982: 141–55.

Rumbaugh, Duane M., and Timothy V. Gill. ''Lana's Acquisition of Language Skills.'' Language Learning by a Chimpanzee: The LANA Project. Ed. Duane M. Rumbaugh. New York: Academic Press, 1977. 165–92.

Savage-Rumbaugh, E. Sue. Ape Language from Conditioned Response to Symbol. New York: Columbia UP, 1986.

——. ''Verbal Behavior at a Procedural Level in the Chimpanzee.'' Journal of the Experimental Analysis of Behavior 41 (1984): 223–50.

Savage-Rumbaugh, E. Sue, and Duane M. Rumbaugh. ''Communication, Language, and Lana: A Perspective.'' Language Learning by a Chimpanzee: The LANA Project. Ed. Duane M. Rumbaugh. New York: Academic Press, 1977. 287–309.

Savage-Rumbaugh, E. Sue, Duane M. Rumbaugh, and Sarah Boysen. ''Do Apes Use Language?'' American Scientist 68:1 (1980): 49–61.

Savage-Rumbaugh, E. Sue, James L. Pate, Janet Lawson, Tom S. Smith, and Steven Rosenbaum. ''Can a Chimpanzee make a Statement?'' Journal of Experimental Psychology: General. 112.4 (1983): 457–92.

Seidenberg, Mark S., and Laura A. Petitto. ''Ape Signing: Problems of Method and Interpretation.'' The Clever Hans Phenomenon: Communication with Horses, Whales, Apes, and People. Annals of the New York Academy of Sciences Volume 364. Ed. Thomas A. Sebeok and Robert Rosenthal. New York: New York Academy of Sciences, 1981. 116-27.

Terrace, H. S. ''How Nim Chimpsky Changed My Mind.'' Psychology Today 1979: 65-76.

——. ''In the Beginning was the 'Name.''' American Psychologist 40 (1985): 1011-28.

——. ''Is Problem-solving Language?'' Rev. of Intelligence in Apes and Men, by David Premack. Journal of the Experimental Analysis of Behavior 31 (1979): 161-75.

——. Nim. New York: Knopf, 1979.

——. ''Why Koko Can't Talk: The Ape's Still Fooling Most of the People Most of the Time.'' The Sciences 22 (1982): 8-10.

Terrace, H. S., L. A. Petitto, R. J. Sanders, and T. G. Bever. ''Can an Ape Create a Sentence?'' Science 206 (1979): 891-902.

Von Glasersfeld, Ernst. ''Linguistic Communication: Theory and Definition.'' Language Learning by a Chimpanzee: The LANA Project. Ed. Duane M. Rumbaugh. New York: Academic Press, 1977. 55-71.

The Process of Writing a Research Paper

The research paper is a long report or a long argument or both. It is generally a longer paper; you are expected to do a thorough job of finding all the evidence on a research topic. However, for a school paper, if you are not given directions about length assume approximately a dozen pages. You must collect a very complete bibliography before doing anything else. A formal research paper requires an outline and an abstract. See Chapter One for an abstract of the ape-language paper.

The introduction to a research paper has a significant job to do. Note that Lisa begins by discussing her research question and citing evidence concerning both the research problems and the significance of her question. Most research papers begin by stating the thesis (see p. 214 for various introductory strategies). It is customary to give the background to the question at the outset; whether you should also discuss significance at the outset, as part of the background, is something for you to decide. Many researchers prefer to reserve significance for the conclusion. Note that Lisa's exact research question comes at the end of her introduction, where it makes a natural transition to the rest of the paper. Whether the thesis statement or question comes first, last, or in the middle of the introduction is up to you. It depends on your audience and purpose and to some extent on the nature of your research. Note that the introduction is not called "introduction"; use a descriptive label instead. (Headings and labels are recommended in APA style; they are optional in MLA style.)

Organize your data for efficient presentation. If you are writing a review of research, simple chronological order may be best. If you are examining a research controversy, an issue with two or more sides, you may need to use order of importance. Lisa is writing a review of research, but her thesis question amounts to a controversy: some researchers believe apes can learn language, some do not. She has begun with the "weaker" side—that side her evidence convinces her is "wrong." In presenting the history this way you must present it as positively and forcefully as possible. This is not the place to find fault with the research. Note that although this research has been seriously criticized, Lisa does not introduce the criticism here.

Before getting to the Critical Views, Lisa ends her presentation of the ape experiments with a summarizing paragraph. This paragraph serves as a transition to the next section. It helps to remind the reader, in capsule form, what the evidence so far has shown. This is not a required paragraph, but anything you do to help your reader follow the discussion is a good idea.

Research papers, by definition, are compilations of information, and the more information the better. Lisa's many reference notes help us, the readers, believe her. That is, your credibility is tied to the number and quality of your references (see Chapter Five).

Since this paper investigates a two-sided issue, Lisa has separated the two sides. In the first half of her paper she has presented the views of those who believe apes can learn language. Her research has caused Lisa to doubt this, but she presents this view without bias. In the second half of her paper she presents the opposing view. The tone of her paper is objective, unbiased. Lisa is teaching us, the readers, what the argument is about—without taking sides. The arguments against ape language are not Lisa's arguments; they are the arguments she has found in the research. Note that she has withheld the most important criticism, Cuing: The Major Flaw, for last.

In the conclusion or Results section of your paper, you should review the key points and evaluate them. The summary ending is a cliché and should be avoided. However, in line with a general policy of doing whatever may help the reader to follow the discussion, it is an excellent idea to start your final section, as Lisa has, with a very brief summation of the points on both sides. Finally, you must answer your research question. Having reviewed the arguments, the answer then follows, and your reader should be convinced that your answer is the correct one. That is, given the evidence as you have presented it, your conclusion should be logical—and inevitable. If you have done a thorough job of researching and documenting, your reader should agree with your conclusion.

There is little documentation in this section; it has all been covered in the body of the paper. However, Lisa has saved something for her conclusion (see Chapter Six). She ends with a final quote that is relevant to her conclusion. Note that this section is not called Conclusion; all section headings should have distinctive, descriptive labels.

If you have content notes, they should follow the ending on a separate Notes page, numbered with raised (half a line) superscripts. Start a new page for your Works Cited or References bibliography. Make sure your bibliography matches your references. Anything in your bibliography must be there because you cited it in your paper. Any reference not mentioned in your paper must not be in your bibliography. Other material which you read but did not actually use in your paper (did not mention by name) should not be included in the bibliography. (See Chapter Five).

ACTIVITY 53 Write a research paper. You are expected to write a thoroughly researched paper with references. Plan for a longer-than-usual paper, including bibliography. Assume a two-sided question: A question of ethics, legality, research accuracy or some other issue has arisen in your area of expertise (for example, should we ban handguns?). In order to reach decisions in this matter, the deciding body must have an accurate, objective report. You have been hired to investigate the issue, look into all sides of the question, read the relevant research, determine the distribution of evidence and number and quality of arguments, and show which, if any, side of the question is best supported by the available data, without expressing any opinions or making any judgmental comments either way.

ACTIVITY 54 As an alternative to the issue-oriented or two-sided research question, you may wish to write a long report. Assume you are a journalist, historian, or researcher with an assignment to write a long report on the history of an idea, the status of a situation, a biography of some historical figure, or a similar report for the appropriate audience.

STUDY GUIDE: Appendix Writing Situations

1. How much diversity and variation is there in research writing?
2. What is the basic principle of research writing?
3. What is the value of summarizing?
4. How long should a summary be?
5. What is a descriptive summary?
6. What is an objective summary?
7. What is an abstract?
8. What is the process of writing a summary?
9. What is the meaning of "to analyze"?
10. What is a rhetorical analysis?
11. What is a pentad analysis?
12. What is the process of writing an analysis?
13. What is the point of a critique?
14. What is the process of writing a critique?
15. What determines the appropriate point of view in a critique?
16. What is the point of writing a comparison?
17. What is the process of writing a comparison?
18. What are the options for organizing a comparison paper?
19. What is a review?
20. What is the difference between objective and critical reviews?
21. What is a balanced critique?
22. What is the process of writing a review?
23. What is a report?
24. What is the process of writing a report?
25. What is argumentative writing?
26. What is the process of writing an argument?
27. What is the appropriate tone of a scientific or academic argument?
28. What are the requirements for a research paper?
29. How should a research paper be organized?
30. How should the Results or conclusion of a research paper be written?

Index